Birdology

"Convincing detail . . . engaging style . . . fascinating facts. . . . Montgomery's stated goal to 'restore both our awe and our connection to these winged aliens' is admirably reached in this book."

　　　　—*Provo (UT) Daily Herald*

"*Birdology* strives to correct the dated but still-recent scientific assessment of animals in general, and birds in particular, as unfeeling automatons. Luckily, science has almost caught up to what every pet owner knows, and Montgomery avails herself of new findings as she brings to life parrots that converse rather than merely 'parrot,' crows that use tools, and pigeons that navigate by making use of everything from Earth's magnetism to smell."

　　　　—*The Washington Post*

"Montgomery gives herself over so wholeheartedly to animals and other humans who share her passion for creatures both rare and ubiquitous that her nature chronicles are uniquely radiant. . . . Expresses profound appreciation for the living web of life in a book that both bird lovers and readers new to bird lore will find evocative, enlightening, and uplifting."

　　　　—*Booklist*

"An original, even brilliant, account of seven species of birds—their fundamental strangeness and their strange familiarity. . . . I have learned something from every chapter."

　　　　—*Living Bird*

"From the first greeting of her charming hens to her celebration of crows in all their smarts and wit, Sy Montgomery has me hooked."

　　　　—Dr. Donald Kroodsma, author of *The Singing Life of Birds*

"Sy Montgomery has once again taken animal writing to another level of consciousness, compassion and understanding. *Birdology* is a brilliant, strikingly original, and beautifully written look into the fragile, rich story of birds, whose lives are more varied, endangered, individualistic, and powerful than dogs and cats. I've never read anything like this about birds. These stories will haunt and enrich you."

　　　　—Jon Katz, author of *A Dog Year* and *Katz on Dogs*

For Adults:

Walking with the Great Apes:
Jane Goodall, Dian Fossey, Biruté Galdikas

Spell of the Tiger:
The Man-Eaters of Sundarbans

The Curious Naturalist:
Nature's Everyday Mysteries

The Wild Out Your Window

Journey of the Pink Dolphins:
An Amazon Quest

Search for the Golden Moon Bear:
Science and Adventure in Pursuit of a New Species

The Good Good Pig:
The Extraordinary Life of Christopher Hogwood

For Children:

The Snake Scientist

The Man-Eating Tigers of Sundarbans

Encantado:
Pink Dolphin of the Amazon

Search for the Golden Moon Bear:
Science and Adventure in the Asian Tropics

The Tarantula Scientist

Quest for the Tree Kangaroo:
An Expedition to the Cloud Forest of New Guinea

Saving the Ghost of the Mountain:
An Expedition Among Snow Leopards in Mongolia

Kakapo Rescue:
Saving the World's Strangest Parrot

Birdology

Adventures with Hip Hop Parrots,
Cantankerous Cassowaries, Crabby Crows,
Peripatetic Pigeons, Hens, Hawks, and
Hummingbirds

Sy Montgomery

Free Press
New York   London   Toronto   Sydney

Again, as always, for Dr. Millmoss

Free Press
A Division of Simon & Schuster, Inc.
1230 Avenue of the Americas
New York, NY 10020

Photography credits are listed after the index.

First Free Press trade paperback edition February 2011

FREE PRESS and colophon are trademarks of Simon & Schuster, Inc.

For information about special discounts for bulk purchases, please contact Simon & Schuster Special Sales at 1-866-506-1949 or business@simonandschuster.com.

The Simon & Schuster Speakers Bureau can bring authors to your live event. For more information or to book an event contact the Simon & Schuster Speakers Bureau at 1-866-248-3049 or visit our website at www.simonspeakers.com.

Book design by Ellen R. Sasahara

Manufactured in the United States of America

3   5   7   9   10   8   6   4   2

The Library of Congress has cataloged the hardcover edition as follows:

Montgomery, Sy.
Birdology : adventures with a pack of hens, a peck of pigeons, cantankerous crows, fierce falcons, hip hop parrots, baby hummingbirds, and one murderously big living dinosaur / Sy Montgomery.
p. cm.
Includes bibliographical references and index.
1. Birds. 2. Birds—Psychology. I. Title.
QL676.M7585 2010
598—dc22   2009031303

ISBN 978-1-4165-6985-5
ISBN 978-1-4165-7016-5 (ebook)

Contents

Contents

Introduction

I have waited a long time to write this book.

Ever since I was privileged to live among and study wild emus in South Australia in 1984, I have wanted to write a book about birds. Though I have written elsewhere about those emus, I never got around to a whole book about birds, mostly because mammals kept jumping the line: chimps, gorillas, and orangutans. Man-eating tigers. Amazon River dolphins. Snow leopards. Tree kangaroos. Our wonderful pig, Christopher Hogwood. Putting off this book was not intentional, but perhaps fortunate. It is easier to understand our fellow mammals; birds are more distant and enigmatic. It has taken me all this time even to begin to learn how to probe their mysteries.

Birds have been trying to educate me since I was a child. I have always loved watching and reading about and learning from them. But even more importantly, birds always come into my life at critical moments to enrich my spirit and enlarge my heart.

The first male ever to court me—I was only a child but I loved him deeply—was a bird. I was seven. His name was Jerry. He was a green parakeet who came home with me from a dime store. And although I was delighted when he first agreed to perch on my hand and thrilled when he would finally fly to me, I was most honored when he threw up on my finger. Even though I had never seen any other creature do this in quite the same way, I understood what was happening: he was feeding me. This was a mark of deep trust and affection, and he would do this only with me. He had taken me as his mate.

I was entranced and honored by everything Jerry did. I loved the way he used his curved beak to hull his round millet seeds. He'd pierce the husk with his lower mandible and peel the outer husk away by forcing the seed against his ridged palate and twirling it with his strong, muscular tongue. Jerry was thrillingly different from everyone else I had known. Even his digestive system was a radical departure from those of mammals, as I could clearly see from his droppings, which appeared from a single opening and incorporated both waste products in one neat little two-colored package. And of course I loved that Jerry could fly. I let him do so often, to my mother's dismay. He liked to perch on the crystal chandelier over her prized mahogany dining room table, with predictable consequences.

People used to ask me if I wanted to teach Jerry to talk. I did not. I already had friends who talked. I wanted him to teach *me* what *he* knew, for I realized that he was the master and I the student. Among the many things he showed me was that birds stir our souls in ways that change our lives.

The ancients knew this well. Images of birds adorn the walls of the caves of Lascaux, accompanying prehistoric humans on the hunt. It's easy to imagine why people carved pictures of birds on tombs as early as 2600 B.C. Birds fly through the air, the element of our breath; they symbolize the soul. From Sufi poetry to native American myths, from Egyptian tombs to Christian tapestries, birds beguile us. With their beauty, their songs, their flight, and their very strangeness, birds stir our deepest psychic strivings. Even the commonest of birds (remember Jonathan Livingston Seagull?) wield the power to thrill and inspire.

No wonder bird-watching is the fastest-growing of all outdoor activities in the United States and one of the most popular hobbies the world around. Happily, there are many hundreds of excellent books in print on bird-watching and classification. This book is not among them. Instead, these pages address a dilemma about which British author Tim Dee writes in his book *A Year on the Wing.* "We seek the truth of birds through collection and classification," he writes. But the truth of birds eludes us. Of a woodcock he recorded at age thirteen, he writes, "I was sure of our identification . . . but it offered so little of itself that I felt I had been simultaneously shown the bird and excluded from it." He calls this "the defining condition of bird-watching." We can know a bird's name; we can identify it and by its sighting add it to our "life list"; but still, the essence of the bird flies away.

Even though birds are all around us, most of them are strangers. We don't know them as individuals. We know very little of what it might be like to be a bird. If we did, we would be awestruck. That is the purpose of this book: to restore both our awe and our connection to these winged aliens who live among us.

Birds are the only wild animals most people see every day. No matter where we live, birds live with us. Too many of us take them for granted. We don't appreciate how very strange they are, how different. We don't realize what otherworldly creatures birds are.

Their hearts look like those of crocodiles. Birds are covered with modified scales—we call them feathers. Their bones are hollow, permeated with extensive air sacs. They have no hands. They give birth to eggs.

No other scientific classification of living creature we commonly see is so different from us as is the class Aves. We don't even think of birds as "animals" (although they are—as are humans, of course). We consider "animals" to be our fellow mammals, with whom our kinship is obvious. It's easy to see a kindred soul when you look into the eyes of a chimp, for instance. They share more than 98 percent of our DNA. You can get a blood transfusion from a chimp. We shared a common ancestor with chimps as recently as 5 million years ago. But actually, all mammals now living (except for the pouched marsupials like kangaroos and the egg-laying platypus and echidna) share at least 90 percent of their genetic material with humans. We shared a common ancestor with even the most distant of our fellow placental mammals as recently as 100 million years ago. The last ancestor we shared with the birds, however, traces back 325 to 350 million years ago.

A bird is as distant from us as a dinosaur. But unlike the extinct monsters of the Jurassic and Cretaceous, birds today are everywhere among us—on our sidewalks, at our bird feeders, on our dinner plates. Yet despite our disparate evolutionary paths, scientists are now beginning to reveal the extent to which birds' emotional and intellectual abilities are remarkably like ours.

To understand birds is to appreciate at once their fundamental strangeness and sameness. What makes a bird a bird? To explore this question, in these pages I share my adventures with seven different bird species. Each chapter explores a different aspect of avian essence. Here are seven essential truths about birds.

In my opinion, the first thing you need to know about birds is that Birds Are Individuals. That's why I begin this book with the Ladies, my flock of emotional, intelligent, and highly individualistic hens, who come when called, love to visit the neighbors, and befriend and feud with fellow chickens. Although a flock of hens is all about community, each chicken is quite distinctive, and the personality of each individual is extremely important to the flock dynamic. People who don't know chickens are always astonished to learn this, but when you are in the company of birds, you must be prepared to be surprised.

A second fundamental truth of birds is that Birds Are Dinosaurs. That may be difficult to see when you're watching a fluffy chickadee at the feeder, but it is abundantly clear when you are crashing through the rain forest of Queensland, Australia, pursuing a 150-pound cassowary, a bird as tall as a man, crowned with a helmet of bone on its head and a killer claw on each foot. Following these birds' tyrannosaur-like footprints, I traveled back in time, to an era of wondrous transformation.

The dinosaurian lineage that became the birds left the earth for the skies. And in doing this, they utterly reshaped their bodies inside and out. We mammals are made mainly of heavy fluids, but as I point out in another chapter, Birds Are Made of Air. Their bones are hollow; their feathers weigh more than the skeleton. Their bodies are full of air sacs; their feathers, also hollow shafted, are sculpted to capture and move air. Birds are essentially feather-fringed bubbles—a fact frustratingly poignant to bird rehabilitators, whose Herculean task it is to nurse back to health, or raise to adulthood, beings whose essential fragility gives them the power to conquer the sky. It was my great luck to apprentice myself to a woman whose job was yet more difficult than that of most bird rehabilitators. Brenda Sherburn's specialty is raising orphaned baby hummingbirds, who hatch from eggs the size of navy beans and are born the size of bumblebees. They are air wrapped in light— the lightest birds in the sky.

What else must we know before we begin to understand what makes a bird a bird? Birds' wildness, for one. Even a pet parakeet, even a barnyard chicken is wild in a way that most mammals are not. To explore this, I embarked on a study of falconry, for the hawks and falcons exhibit a genius for instinct that showcases the brilliance of all birds' wildness.

Birds are able to apprehend the world in ways that we cannot. They can see polarized and ultraviolet light. They experience colors we can never know. They sense the earth's magnetic field, navigate using subtle changes in odor and barometric pressure. They imbibe realities of this world that we cannot fathom and use them to circumnavigate the globe. We are only now starting to understand how birds accomplish these extraordinary feats, by way of one of our most ordinary and unappreciated birds, the pigeon—the hero of yet another chapter.

Though gifted with instincts and senses that we lack, birds' intellectual capacities are shockingly similar to our own. Some birds appreciate human art to the extent that they can learn to tell the difference between the paintings of Monet and those of Manet. Some birds love to dance—and in the course of researching this book, I got to dance with one of them. Birds' capacity for song is of course so legendary that many cultures tell us the birds taught music to humans. There are birds who can even speak to us meaningfully in our own language—something that, many scientists believe, even our close hominid cousins, the Neanderthals, probably could not do. But Birds Can Talk. In these pages, I relate some of what they tell us.

And finally, to apprehend the truth of birds, we need to appreciate birds' ubiquity: Birds Are Everywhere. The final chapter takes us to a winter crow roost of tens of thousands of birds, a congregation that had been traditionally based in the countryside but, like the majority of the world's population of humans, has moved into the city. How do the city's people react to their new urban neighbors? And what does this say about the future of humans and birds living together on this sweet, green earth?

Just as I was beginning to research this book, my friend Gretchen Vogel called me with an unusual suggestion: that we should go to church together. I was raised Methodist, she was raised Catholic, but the church she had in mind was the First Universalist Church in West Chesterfield, New Hampshire, about an hour's drive from my house and twenty minutes from hers.

She had read a notice in our local paper about that week's sermon. The title was "Birdology." How could I resist?

The small church, built in 1830, had no cross in the front, and the ceiling inside was painted a light blue, like the sky. Gretchen and I joined six other people in the congregation that Sunday. Presiding was a visiting pastor, Rev. Elaine Bomford.

A smiling, articulate woman in her fifties, the pastor greeted us with the confession that "birdology" was a word she had made up. She invented "birdologist" to describe the person who—unlike the ornithologist, with his fancy university degree—hasn't completed a formal course of avian study. To be a birdologist, she said, "you just have to appreciate birds and be intentional about appreciating birds in some way." We could all be birdologists, she said—and we should be. For watching birds, she told us, "strengthens our souls."

A birdologist, she explained, "experiences the divinity of creation revealed in the birds."

That pretty much describes what I have always felt when I see a bird— any bird. The ancients believed that birds could bring us messages from the gods: whether a battle would be won or lost, or if plenty or famine would befall the city. Birds do bring us messages from the divine, but not necessarily those the ancients sought. Because of birds' very different lineage; because they are made of different stuff than we; because of the powers that birds possess that we do not; and because, despite our deep differences, we can share much of a bird's mental and emotional experience, birds bring us news far more important than our own personal, human lives. They bring us news about the larger and more wondrous *life*, about a world that we, with our merely human senses, have barely begun to perceive.

Birds teach us reverence—a virtue that, writes classicist and philosopher Paul Woodruff, "begins in a deep understanding of human limitations." No beings show us our limitations better than do the birds. When we see a bird in flight or let our hearts soar on the notes of its song, the mystery of the world wells up before us—a mystery we long to embrace rather than conquer.

That day in the church, Rev. Bomford asked us to speak words with her from Terry Tempest Williams's book *Refuge*.

"I pray to the birds," we read, "because I believe they will carry the messages of my heart upward . . ."

Oh, yes, I thought. It was then the time of the fall migration, and with every overhead "V" I utter a prayer for the birds' safe passage.

"I pray to the birds," we continued, "because they remind me of what I love rather than what I fear . . ."

I thought back to my time with the emus. At first sight, I had fallen wildly, passionately, soul splittingly in love. I had been working alone, collecting plant samples on a wombat preserve in the outback, when I looked up and saw them: three birds standing tall as a man, approaching me on their long, balletic, backward-bending legs—legs so strong they were capable of ending my life with a single kick. Instantly, I was smitten by both their strangeness and familiarity. There was no room for fear. My heart was filled with awe.

Reading Terry Tempest Williams's words, I gave thanks for all the many birds I have loved in my life: my parakeet Jerry; a cockatiel named Kokopelli who liked to sit on my head and whistle the National Geographic theme song when I was on the phone; my beloved Ladies. I did not know then, of course, that this book would bring me many more individual birds to love: the dancing cockatoo Snowball; Harris's hawks named Jazz and Fire and Smoke; a talking African grey parrot named Griffin; two orphaned baby hummingbirds who, God willing, could be raising babies of their own somewhere in California as I write these words right now.

" . . . And at the end of my prayers," the few of us gathered in the church that morning read, birds "teach me how to listen."

This was my prayer as I began working on this book. It's my hope that the birds you will meet in these pages will teach you, too.

—Sy Montgomery
 May 19, 2009
 Hancock, N.H.

Chickens

Birds Are Individuals

H ello, Ladies!"
Even if there is no one in sight, I call out to them whenever I round the corner of the woodpile to enter the barnyard. For even if they're scattered over several acres of lawn and woods and brush—some hunting in the compost pile, others patrolling the neighbor's blueberry patch, some scratching in the leaves by the stone wall—I know they'll come running. A dozen foot-tall, black and black-and-white figures, holding their wings out like tots spreading their arms to keep balance or beating their wings to propel themselves even faster, come racing toward me on scaly, four-toed feet, showing the wild enthusiasm of fans catching sight of a rock star. It's a welcome that makes me feel as popular as the Beatles—even if my personal fan club is composed entirely of poultry.

At times, I suppose, I am less a celebrity than the moral equivalent of the neighborhood ice cream truck. For often, I come bearing food—vegetable peelings from the house, the trimmings from pie crust dough, and sometimes an entire tub of fresh cottage cheese that I buy just for them. My hens, like many pets, particularly enjoy being fed by hand. The lead chicken, standing before me front and center, tilts her head to examine my offering with one skeptical orange eye. Then she seizes the first morsel in her hard black-and-amber beak—and the crowd goes wild. Everyone pecks with great enthusiasm, hard enough to hurt my palm. If, among the buffet, there is one particularly big treat to be had—a single apple core, a baked squash skin—at some point, somebody will seize this. The victor will run some distance, chased by her sisters, until the prize is either stolen or swallowed. This is usually good for about ten minutes of entertainment.

But often, I don't come bearing food. I come just for a visit. I relish these encounters even more. The Ladies don't seem disappointed at all. They mill at my feet, cheerful and excited, for they know I have a different treat in store. They are waiting for me to pick them up, stroke them, and kiss their warm, red, rubbery combs.

They also like me to run my hand along the sleek length of their backs. Each will squat, wings slightly raised, neck feathers erected, welcoming my caress. I start at the back of the neck, and when my hand has completed half its journey, the hen will arch her back. I gently close my fingers around her tail feathers until my stroke swoops into the air—rather like the way you would stroke a cat. Then it starts all over again, until the hen has had enough and has reached what we call "overpet." She fluffs her feathers, shakes, and, fortified by affection, strolls off to continue her chicken day.

When I crouch to pet one hen, another one might hop up to perch on my thigh, patiently waiting her turn. I talk to them. "Hello, Ladies! How are my Ladies? Did you find good worms today? What was in the compost?" They keep up their end of the conversation with their lilting chicken voices.

Visitors who witness this for the first time are amazed. "I've never seen anything like that!" they say. "I always thought chickens were stupid! Is it possible," they wonder, "that they actually know you?"

Of course they know me. They know the neighbors, too. In the more than two decades that I've been living with chickens, they have formed deep bonds with some—and not with others. Certain individual chickens adored our pig, Christopher Hogwood, who lived for fourteen years in the pen next to the coop in the barn. Some even chose to roost with him, perched atop his great prone bulk, instead of spending the night with flock-mates. But none of the hens has ever bonded with our border collies. They never visit the neighbors across the street—but they adore the retired couple next door. When the Ladies hear Bobbie and Jarvis Coffin's screen door slam, the hens hop over the low stone wall separating our yards and rush to greet them. When Jarvis relaxes in the backyard hammock on summer days, they gather beneath him, and some leap into the air, attempting to join him in his day roost. (So far they haven't succeeded.) The hens mob the couple whenever they try to enter their cars. Usually Bobbie and Jarvis get them some cracked corn, which they keep in their shed just for our hens, and make their getaway while the birds are eating. Sometimes, Bobbie confesses, when she's in a hurry, she sneaks out the door and tiptoes to the car, to avoid a protracted visit with our chickens.

Sometimes, the Ladies don't wait for the Coffins to make the first move. A few of the bolder hens have been known to mount the flight of wooden steps leading to Bobbie and Jarvis's second-storey back porch—quite a feat considering the birds are only twice as tall as the steps are deep. They gather, softly discussing their plans, outside the porch door, looking in through the glass panes, trying to catch the attention of their human friends and entice them to come out and play.

Jarvis Coffin with the Ladies

Occasionally the hens come over while the Coffins are hosting a gathering. Their guests are invariably impressed. "I didn't know you had chickens!" people exclaim—and then seem dumbfounded that, like themselves, our chickens simply enjoy visiting their lively, kind neighbors.

Folks use words like "astonishing" to describe such friendships between people and poultry. What's more astonishing, though, is not that these birds know so much about their human neighbors, but that we humans know so little about our neighborhood birds—even one as common and readily recognized as a chicken.

People think they know lots about chickens, and you'd think they would: there are about 20 percent more chickens on earth than people, and since by definition they are domestic fowl (they are a separate subspecies from their wild ancestor, the red jungle fowl of Southeast Asia), all of them live among humans. There are at least four places in the United States named after chickens, including the towns of Chicken, Alaska, and Chickentown, Pennsylvania; in 1939, the Delaware state legislature, braving the opposition of the State Federation of Women's Clubs of Delaware (which wanted the cardinal instead), selected the Blue Hen Chicken, a local variety but not a true breed, as its state bird.

Still, chickens are rarely celebrated in our culture and rarely given the respect they deserve. I once sat next to a man on an airplane who detailed for me at length the attributes of the species: they are stupid, disgusting, filthy, cowardly, occasionally cannibalistic automatons, he said. How had he acquired this opinion? It turned out he had worked at a factory farm—the sort of place where most chickens are raised for food in the United States— in a dirty, overcrowded warehouse that resembled a prison camp.

This is not the best place to get to know someone. Nor is a dinner plate. Yet for most of us, our relationship with chickens is generally of a culinary nature. In fact, the first definition for the word "chicken" I encountered on the Web doesn't even mention that it's a bird. It's "the flesh of a chicken used for food." The average American eats more than eighty pounds of chicken per year, according to the National Chicken Council, making it the most popular meat consumed in the United States. Some 8.9 billion birds yearly are roasted, boiled, Kentucky Fried, and turned into everything from McNuggets to the famous "Jewish penicillin," chicken soup. (One of my friends, an award-winning journalist and talented cook, was shocked to learn, when she was nearly forty, that it is physically possible to make soup without chicken stock.)

The disturbing fact that, on the way to the soup pot, a chicken can continue to run around after decapitation does little to bolster appreciation for the species' more refined traits. In fact, one rooster was able to live for eighteen months after his head was cut off. Farmer Lloyd Olsen, hoping to please his visiting mother-in-law, who particularly savored boiled chicken neck, failed to kill the rooster when his ax missed the bird's carotid artery and left one ear and most of the brain stem intact. Not only did the victim survive, but he grew from two and a half pounds to eight, and attained national fame as Mike the Headless Wonder Chicken on the side-show circuit from 1945 to 1947. Even now, the rooster's hometown of Fruita, Colorado, holds Mike's Festival the third weekend in May each year—a day of races, games, and food intended, as its organizers say, to celebrate the bird's "admirable will to live." But alas, Mike's story also perpetuates the one "fact" most people claim to know about chickens: that they are automatons too stupid to know if they're dead or alive.

But, as I've come to learn over the past couple of decades sharing my life with successive flocks of these affectionate, industrious, and resourceful birds, almost everything people "know" about chickens is wrong.

My friend Gretchen Vogel began my education. When my husband and I bought the 120-year-old farmhouse we'd been renting in southwest New Hampshire, she gave us our first flock of hens as a barn-warming present. Gretchen raised organic vegetables and Connemara ponies on a farm near our house, and when she had moved there, years earlier, a friend had given her the gift of twelve hand-raised black hens. Gretchen had been delighted at the prospect of fresh free-range brown eggs every day, but the flock provided much more. "It was the most incredible gift anyone ever gave me," she said. "I had been given the gift of an entire world—a whole chicken universe."

Of course, back then, I had no idea what she was talking about. But I was eager for a flock of my own. Every animal I have ever known has bettered my life in some way. I love birds—I have lived with finches and parakeets, cockatiels and lovebirds, and a gorgeous red and blue Australian parrot called a crimson rosella—but I had never lived with a whole flock of them. What would the hens reveal? "You'll see," Gretchen promised. "Nothing has ever made me happier."

At the Agway feed store, Gretchen ordered for us twelve chicks of the same breed she'd first owned—Black Sex Links, so named because the females can be identified upon hatching by their all-black color, averting the problem of raising a coop full of jealous roosters—and hand-raised them in a heated trailer on the farm. My husband and I would often visit them there, holding one or two peeping chicks in our hands, or on our laps, or tucked into our sweaters, speaking softly to each so she would know us. When they were old enough—no longer balls of fluff but sleek, slim black miniatures of their eventual adult selves—they moved into our barn. Our travels in the Chicken Universe had begun.

At first I was afraid they'd run away or become lost. We had a cozy, secure home for them prepared in the bottom storey of our barn, with wood shavings scattered over the dirt floor, a dispenser for fresh water, a trough for chick feed, some low perches made from dowels, and a hay-lined nest box made from an old rabbit hutch left over from one of the barn's previous denizens, in which they could lay future eggs. Chickens need to be closed in safe at night to protect them from predators, but by day we didn't want to confine them; we wanted to give them free run of the yard. But how could they possibly understand that they lived here now? Once we let them out, would they even recognize their space in the barn and go back in it? When I was in seventh grade, my family had moved, once again, to a new house; my first afternoon there I literally got lost in my own backyard. Could these six-week-old chicks be expected to know better?

Gretchen assured me there would be no problem. "Leave them in the pen for twenty-four hours," she told me. "Then you can let them out and they'll stick around. They'll go back in again when it starts to get dark."

"But how do they know?" I asked.

"They just do," she said. "Chickens just know these things."

When before dusk, I found them all perched calmly back in their coop, I saw that Gretchen was right.

In fact, chickens know many things, some from the moment they are born. Like all members of the order in which they are classified, the Galliformes, or game birds, just-hatched baby chickens are astonishingly mature and mobile, able to walk, peck, and run only hours after leaving the egg.

This developmental strategy is called precocial. Like its opposite, the altricial strategy (employed by creatures such as humans and songbirds, who are born naked and helpless), the precocial strategy was sculpted by eons of adaptation to food and predators. If your nest is on the ground, as most game birds' are, it's a good idea to get your babies out of there as quickly as possible before someone comes to eat them. So newborn game birds hatch covered in down, eyes open, and leave the nest within twenty-four hours. (An Australian game bird known as the malleefowl begins its life by digging its way out of its nest of decaying vegetation and walks off into the bush— without ever even meeting either parent.)

That chickens hatch from the egg knowing how to walk, run, peck, and scratch has an odd consequence: many people take this as further evidence they are stupid. But instinct is not stupidity. (After all, Einstein was born knowing how to suckle.) Nor does instinct preclude learning. Unlike my disoriented seventh-grade self (and I have not improved much since), young chickens have a great capacity for spatial learning. In scientific experiments, researchers have trained days-old chicks to find hidden food using both distant and nearby landmarks as cues. Italian researchers demonstrated that at the tender age of fifteen days, after just a week's training to find hidden food in the middle of their cage, chicks can correctly calculate the center of a given environment—even in the absence of distinctive landmarks. Even more astonishing, they can do it in spaces they have never seen before, whether the area be circular, square, or triangular. How? The chicks "probably relied on a visual estimate of these distances from their actual positions," wrote University of Padova researcher L. Tommasi and co-authors in the *Journal of Comparative Physiology,* " . . . [but] it remains to be determined how the chicks actually measure distances in the task."

We never determined how our first chickens knew their new home was theirs, either. We never knew how they managed to discern the boundaries of our property. But they did. At first, they liked to stay near the coop. But as they grew, they took to following me everywhere, first cheeping like the tinkling of little bells, later clucking in animated adult discussion. If I was hanging out the laundry, they would check what was in the laundry basket. If I was weeding a flower bed, they would join me, raking the soil with their strong, scaly feet, then stepping backward to see what was revealed. (Whenever I worked with soil, I suspect they assumed I was digging for worms.) When my husband, Howard, and I would eat at the picnic table under the big silver maple, the Ladies would accompany us. When my father-in-law came to help my husband build a pen for Christopher Hogwood, then still a piglet, the Ladies milled underfoot to supervise every move. The hens were clearly interested in the project, pecking at the shiny nails, standing tall to better observe the use of tools, clucking a running commentary all the while. Before this experience, Howard's dad would have been the first to say that he didn't think chickens were that smart. But they changed his mind. After a few hours I noticed he began to address them. Picking up a hammer they were examining, he might say, directly and respectfully, "Pardon me, Ladies"—as if he were speaking to my mother-in-law and me when we got in the way.

But when their human friends are inside, and this is much of the time, the Ladies explore on their own. A chicken can move as fast as nine miles an hour, which can take you pretty far, and ours are free to go anywhere they like. But ours have intuited our property lines and confine their travels to its boundaries. They have never crossed the street. And for years, they never hopped across the low stone wall separating our land from that of our closest neighbor. That came later—and it was not the result of any physical change in the landscape, but the outcome of a change in social relationships among their human friends.

When the Ladies first moved in with us, Larry Thompson lived next door with his Airedale, Cooper, both of whom we liked and visited. When he moved out, the house sat vacant for a time. Still, the hens wouldn't venture over the low wall. Finally Lilla Cabot and her two blond, blue-eyed girls, Jane, seven, and Kate, ten, moved in. Understandably enchanted with our friendly black-and-white spotted pig, the girls visited the barn regularly, bringing treats (often their school lunch), petting him, and escorting him on daily rooting excursions. Next the girls were helping me gather the Ladies' eggs and tossing the Frisbee for Tess, the border collie. Soon we were together baking cookies, reading animal stories, and visiting back and forth daily. That's when the Ladies started hopping over the stone wall. Somehow, they realized, before we humans did, that our two families had become one unit.

This should not have surprised me. To chickens, relationships are extremely important. Researchers have documented that an average chicken can recognize and remember more than one hundred other chickens. How? They may well remember a gestalt of features, including the voice. But facial features seem to be particularly important. Researchers A. M. Guhl and L. L. Ortmann fastened fake combs on hens' heads, to discover the new headgear rendered them strangers to their flock-mates. So did dyeing the feathers on the head. Dyeing the feathers new colors elsewhere on the body or even removing patches of feathers did not. Like us, birds seem to look into the faces of their friends. In his revelatory *The Minds of Birds,* Alexander Skutch tells the story of a small bird of prey known as a kite, who had been fed by a particular soldier, Derek Goodwin, at an army camp in Egypt during World War II. Soaring above marching columns of identically dressed soldiers, the bird would find and hover above Goodwin and Goodwin alone when Goodwin looked up, revealing his face.

Because chickens live in flocks, the ability to identify individuals is even more important than it is to a kite. Belonging is essential to a chicken's well-being, as is clear from the complex social system of the pecking order. The pecking order is not always a straight hierarchy. In a study of captive red jungle fowl, chickens' immediate ancestors, a University of Florida researcher constructed "sociograms" diagramming the relationships between all the birds in each of four flocks. In many flocks, there were several groups of up to three hens who were not only friends but co-equals.

The pecking order is more about order than pecking. Chickens do peck—sometimes to the death—but mine, over the course of many years and many flocks, have pecked with admirable discretion and restraint; sometimes an "air peck" or merely raising the hackles gets the message across. Such a gesture is delivered to remind another chicken of her position in the group. It's important that everyone knows her place. When it comes to roosting at night, the pecking order determines who sleeps next to whom, and on which perch. It does not always determine who eats first, but usually predicts, if there is a squabble between individuals over a choice food morsel, who will ultimately win.

Within this well-established order, ours is a peaceful flock. But it is not immune to violence. A skunk (whom we later captured, moved, and released) dug in through the dirt floor on subsequent nights and killed and ate two hens. A fox carried off another. A wandering dog killed more. The little flock was shrinking. Though far longer than the mere five to eight weeks of life of the supermarket chicken, alas, the average natural life of a pet hen spans only five or six years (though Matilda, an ivory-colored Red Pyle bantam who lived in Birmingham, Alabama, lived to sixteen—earning her an entry in *Guinness World Records* and a spot on *The Tonight Show*). I thought sadly of the day my flock would be reduced to a handful of ancient, menopausal hens. But again, Gretchen knew what to do, and I do this every few years: augment the flock by adding new babies I hand-raise myself.

They come in the mail—like a fruit-of-the-month order, or a book from Amazon.com. But this package is peeping when it arrives. I always tell everyone at the post office to watch out for my special delivery, and Mike or Janet calls me the moment the order arrives from Cackle Hatchery in Lebanon, Missouri—a box of live baby chicks, just hatched two days before.

Lovingly I lift the perforated lid to a straw-lined cardboard box, not much larger than a big box of chocolates. The fluffy, peeping babies are still shaped like eggs. They'll never see the hen who laid the eggs from which they hatched. I'm their mother now, and I love them with a fierce tenderness that never abates.

In my home office where I write, I give the chicks free run of a big box that once shipped a refrigerator, carpeted with newspaper and wood shavings and warmed by a heat lamp. From the first time I did it, raising chicks in my office seemed perfectly normal to me; I showered in the morning with a cockatiel, slept with a dog, and spent many sunny summer hours lying in a field with a pig. Why shouldn't I have peeping chicks in my office?

Most of the day, at least one chick, often two, is somewhere on my body.

For the next six weeks, until their baby down is replaced with feathers, I spend my days writing with a chick or two on my lap, beneath my sweater, on my shoulder or knee. Yanking at the tiny gold cross around my neck, or, worse, hopping onto the keyboard signals time to switch chicks. When I speak with strangers on the phone, they've been known to ask, "Are you calling from a zoo?"

Meanwhile, something magical has happened. Konrad Lorenz, the Nobel laureate credited with founding the modern study of animal behavior, called it imprinting. Students of animal behavior are careful to note that imprinting is not instinct; it is not learning; it is something else entirely. Most newly hatched game birds, including turkeys, ducks, chickens, and geese, will follow the first moving object they see, which is usually, of course, their mother. But working with hatchling greylag geese, Lorenz discovered that the babies don't instinctively recognize adult geese as members of their species. If a person is the first moving creature the gosling sees, the baby will follow the person as if he were the parent—as attested by many a charming photo of the white-bearded scientist walking down a path or rowing across a pond in his hometown of Altenberg, Austria, with a string of fluffy goslings following in single file behind.

Raising baby chicks in my office involves a great deal of cleanup. Unsightly blotches stain my clothing and dry in my hair. And when the downy chicks begin to grow feathers, every surface in my office—my books, photos, maps, notebooks, computer—is coated with a thick layer of powdery dust. (Each new feather grows in as a quill coated with a keratin sheath. As the feather blossoms, the keratin breaks off in tiny pieces. Though this will happen again whenever a feather is replaced, never again is dust released in such quantities.) But it's well worth it. I adore these fearless, busy little souls, already so full of life and purpose. I am honored to follow in the footsteps of the great Lorenz—as my chicks will follow in mine. And imprinting has a later benefit as well: human-imprinted babies later direct toward their person many of the inborn responses that normally would be shared only with a member of their own species. In this way, I become an honorary chicken.

* * *

Chickens with poufy topknots; chickens with feathered feet; chickens with turquoise earlobes; chickens who lay green eggs (though not with ham); chickens with tails that can grow twenty feet long . . . When I placed my initial order with the hatchery, I had many breeds to choose from. Chickens have been living with people for a very long time (by some reckoning, as long as eight thousand years—longer than donkeys and horses, longer than camels or ducks, and by some accounts, even longer than pigs and cattle). Starting with the red jungle fowl of Southeast Asia (who looks pretty much like the rooster on the Kellogg's Corn Flakes box), through selective breeding people have created as many as 350 different varieties of chickens— chickens spangled with iridescent feathers, chickens with naked necks (these are called turkens but are really chickens, not crosses with turkeys), chickens standing tall on long legs like basketball players, miniature chickens called bantams who might weigh only a pound.

Every winter we muse over the catalogs of the chicken hatcheries the way gardeners dream over seed catalogs. We glance at the bargains: one catalog carries a Top Hat Special, an assortment of crested breeds from the five-toed French Black Mottled Houdans to the golden Buff Laced Polish, all of whom look like they are wearing giant Afro wigs made out of feathers. Even the babies sport little top hats. There's usually a special on Feather Footed Fancies: all these varieties have feathered feet, making the birds look a bit like they are wearing floppy slippers. There is even a Fly-Tyer's Special, a selection of chickens whose feathers can be used to fashion particularly attractive fishing lures. Or course, we ignore the Frying Pan and Barbeque specials; I'm a vegetarian and certainly wouldn't eat anybody I know.

What we're really looking for are handsome, vigorous chickens who do well in cold climes. With their glossy black feathers, red, upright combs, and ample bodies, our Black Sex Links were all these things. We sometimes called them "the Nuns," especially when they raced out of their coop each morning like a flock of chatty Sisters leaving a convent in their billowy black habits. Such a uniform appearance did they present that, without noting the subtle differences in the shape of their combs, it was impossible for us to tell them apart. (Scientists faced with this problem sometimes outfit the hens with numbered armbands affixed to the wing.)

Adding birds of different breeds presaged an important change in our understanding: now that it was easier to tell birds apart, the distinct person-alities of individuals began to reveal themselves more clearly.

Kate and Jane next door took this opportunity to give the chickens names. I had not done so before because of a military adage learned from my father, an army general, which warned, "Never name the chickens." I knew this wasn't about poultry, but about the commander's responsibility to remain objective about the troops he must choose to send into battle. But somehow it always stuck with me that something bad would happen if you named your hens. In fact, the only one of the Nuns who'd been named before the arrival of the new babies bore the badge of near-catastrophe: the girls named her Foxy Lady after an encounter with a fox left her with no tail feathers. But with the coming of age of our creamy buff Speckled Sussex and our black and white Lakenvelders came more names: Snow White for one of the "Lakes," who was exceptionally beautiful and loved to fly; Madonna for a loud and theatrical Speckled Sussex; Matilda for one of the remaining Black Sex Links, who, now older, walked with a rolling gait resembling a waltz . . .

Soon it became evident that some hens were consistently outgoing and others shy; some were loud and others quiet; some cautious and others reck-less. This was particularly obvious whenever the hens faced a threat, such as a hawk flying overhead. Some hens hid in pricker bushes; others raced inside the coop. Some dashed behind a large board that leaned against an outside wall of the barn. Some individuals would continue to hide for more than an hour. They were so good at hiding that sometimes, alerted by their alarm calls, I'd rush outside to try to defend them and spend half an hour trying to find them and entice them from their hiding places. Brassy Madonna was often first to emerge, while Foxy Lady, surely remembering the horrible fox, stayed immobile the longest. But all of them understood that even though the hawk might not be visible, it still might be lurking somewhere nearby.

Chickens both remember the past and anticipate the future. This has been clearly demonstrated in the laboratory. In one study, published in the journal *Animal Behavior,* Silsoe Research Institute biologist Siobhan Abeyesinghe and her co-authors tested hens with colored buttons. When hens pecked at a particular button they were rewarded with food. But they got an even bigger reward if they learned to postpone their pecking. If they waited—for up to twenty-two seconds—they got even more to eat. The birds chose to wait for the jackpot more than 90 percent of the time.

Experiments like this show that chickens "can do things that people didn't think they could do," said Christine Nicol, professor of veterinary science at the University of Bristol in England. "There are hidden depths to chickens, definitely."

In our attempt to plumb those depths, the girls and I tried to decipher the chickens' language. At first my husband dismissed our efforts, insisting that most of what they were saying was, "I'm a chicken. You're a chicken. I'm a chicken." He gave them more credit than most scientists did for many years. Even though birds have the greatest sound-producing capabilities of any vertebrate—far superior in both volume and range to the greatest human opera star—their distinctive calls and elaborate songs were not considered true communication. Even parrots who spoke sensible phrases in English to their human owners were dismissed as mere mimics. Birds' spectacular voices were merely unconscious, uncontrolled noises reflecting the birds' inner states (which were also assumed to be unconscious).

The ancients knew better. The word "augury" comes from the Greek word meaning "bird talk," for to understand the language of birds was to understand the gods. And the Cabot girls and I knew better, too. We could feel the anguish in the Ladies' calls when they spotted a predator; we could read their delight when someone found a mother lode of worms or beetles in the compost pile. We discovered, too, that some hens announce the blessed moment when they have laid an egg: a loud, measured series of rising "buk-buk-buk-AHH!"s. We suspected this meant more than just "Ouch!" Hens may perceive their eggs as gifts that may be presented to their rooster, their flock-mates, or an honorary chicken/person. On Farm Life Forum's Web page for poultry keepers, a woman wrote of a chicken her father had kept as a pet when he was a boy. Each evening, the hen appeared at the door of the house and would peck to be let inside. When the door opened, she would proceed directly to the boy's bed—where she would lay an egg on the pillow. Then, gift delivered, she would stride back to the door and return to the henhouse.

At Macquarie University in Sydney, Australia, working with Golden Sebright chickens, a breed whose voices are most similar to the ancestral

jungle fowl, psychology professor Chris Evans and his wife Linda have identified twenty-four separate calls the birds use to communicate specific information to others in the flock. For instance, playbacks of a rooster's kissing
"took-took-took" call caused hens to search for food—evidence it means
"Come, here's some food" and not merely "I'm happy." Critics countered
that hens hearing the call were only succumbing to a knee-jerk reaction: the
call was a trigger causing hens to peck the ground mindlessly. To test this,
the researchers divided the hens into two groups: one got a snack just before
hearing the food call. The others had none. The hens responded as humans
would to language. Those who had just eaten showed limited interest, but
those who were hungry searched the ground for food. "If you're on a long
drive and you pass a restaurant sign, that could be a salient piece of information. But if, after food has been brought to the table, someone says, 'There's
food,' that's a redundant comment. It's that kind of contrast," Chris Evans
explained.

Not only do hens understand when a call is about food; they can even
discern from a rooster's call what's on the menu. The researchers reported the
rooster called at a faster rate if the food discovered is especially tasty—like
their favorite, corn, instead of the regular layer mash ration.

The Evanses also found that chickens used several different alarm calls,
depending on the size, shape, speed, and location of the predator. The
researchers mounted a video monitor in the chickens' cage on which they
could project the images of various predators in different conditions. A
video of a hawk prompted a high-pitched scream, delivered while the bird
crouched. A video of a raccoon elicited pulsating series of ten high-pitched
"buk"s followed by an alarmed "AH!" while the bird paced about. When
these calls were recorded and played to other chickens, the others clearly
understood what they meant: the high-pitched scream made them scan the
sky, while the agitated clucking prompted a search of the ground. The alarm
calls were more vehement when the predator was nearby or approaching
quickly—and in the case of the hawk, delivered more frequently when the
rooster knew he had an audience. (Apparently, a ground-predator call is not
only intended to warn hens but is addressed to the predator as well—probably to let the predator know he's been seen.)

Though these warning calls perhaps best demonstrate how chickens transmit information, the sounds I love most are spoken in sleepy voices, as the Ladies get ready to roost for the night. I check on them each evening and turn out their light. In summer, when the sun sets late, hens are often still milling around the floor. I cry, "Perching!" and in response they fly to their regular roosting spots, each bird surrounded by her closest friends. They settle in for the night, making their long, low, contented nighttime chatter— agreeing, no doubt, that in the Chicken Universe, all is right with the world.

It is easy to believe them. When hens are calm, nothing is more soothing than their voices, especially when punctuated by the occasional grunt of a sleepy pig. Sometimes, lulled by their cozy, restful sounds, I lose track of time, enveloped by a sense of belonging, washed in peace and wholeness among a sisterhood of hens. Some evenings my husband finds me in the henhouse, caressing one or two chickens, eye level with my perching friends, as if one of the flock. He has sometimes overheard me join their evening conversation. "Yes, Ladies," he heard me say one night, "you're my beauties. I love you so much."

Some of the most memorable of the many quirky chickens we've known have been roosters. Only once did we actually order them: seduced by the promise of their glorious long tails, we ordered and paid for two lovely Lakenvelder cockerels.

Our other roosters, though, arrived unbidden. In my earlier days of chicken husbandry, I ordered from a hatchery that, as a bonus, included in every order a "free exotic chick." It might be an Araucana, the kind that lays green eggs, or a stately Blue Andalusian, with its uptight posture and blue feathers. It might be an Egyptian Fayoumi of the Nile, or perhaps a Cuckoo Maran, a French breed laying chocolate-colored eggs . . . It was always exciting to see who turned up in the order. We'd have to wait until the feathers came in to identify the breed with which we had been blessed.

Invariably, though, it would turn out to be a rooster. Always he turned out to be quite handsome; and usually, one day, he would turn on us.

We had heard about this problem. A neighbor's boy had a wonderful rooster as a pet. He used to ride on the handlebars of the kid's bicycle. But then one day, he turned. He attacked his former buddy relentlessly, flying at the boy's face with his spurs. The little boy was bruised and bloodied every day. The parents kept the child, but gave the rooster away.

We felt sorry for the family, but we considered the incident a fluke. Surely nothing like this could happen to us—not with all our chicks firmly imprinted.

Besides, we were thrilled to have roosters. We didn't need a rooster to get our hens to lay (though only fertilized eggs will hatch), but a rooster has much to offer a flock. Hens can hope for no better protector than a good rooster. We can't be in the yard with them every minute, but a rooster can, and he will fight to the death to protect his flock. Flinging himself spurs first at his opponent, a cock will fight with such ferocity and determination that the ancient Greeks believed even a lion would fear him.

Most roosters are very solicitous of their hens. When he's not patrolling for predators, he's often searching for food his flock might enjoy. When he finds it, uttering the food call that the Evanses studied, he stands aside while his women enjoy the treat, and only after they've had their fill will he sample the snack. The Talmud praises the rooster, and its writers advise Jews to learn from him courtesy toward their mates.

We eagerly awaited our cockerels' transition to maturity. When our roosters began to crow, we loved it—once we figured out what was happening. One day I heard an unearthly racket, a sort of strangled gargling, coming from the backyard and rushed outside fearing I'd find a child or some small animal who had been injured. There was our rooster, Clarence, on the stone wall, standing tall and proudly practicing his crow. Like an adolescent singing, it takes a rooster some time to find his mature voice.

Luckily, our neighbors didn't mind the crowing. The Cabot girls and their mom were delighted, and almost everyone else on the street had kept chickens of their own at one time or another. The crow of a cock is a part of the soundtrack of rural life. In the sacred book the Hadith, the prophet Muhammad tells us why roosters crow: they do so because they have seen an angel. The moment a cock crows, the holy man advises, is a good time to ask for God's blessing.

And so is the first time your rooster goes on the attack.

We were utterly unprepared. Our first roosters, the two Lakenvelders, had been such gentlemen, keeping largely to themselves, tending their small flock of seven Lakenvelder hens, mostly ignoring the others. But when our first "free exotic chick" turned into a rooster, that was another story.

Alex the Araucana had been a bold, cheerful chick. A born leader, he would help me round up the other chicks if I needed to close up the coop before dark. Because he was bigger and bossier than the others, we suspected he might be a rooster from the start and thought he would make a fine one. When he started to crow, we knew. But we didn't know yet what darkness lurked in his rooster heart.

One day, Howard was lying on the ground, trying once again to fix our ancient, secondhand lawn mower. From the inches between the machine's steel belly and the soft green grass below, Howard caught sight of something moving fast—malevolently fast—directly at him. Realizing that lying prone on the ground is not a good way to face an attacker, Howard leapt to his feet. Alex pulled up short right in front of him, as if he had come to his senses. "But I knew," said Howard, "that he was up to no good."

We hoped this was just a phase. That it was not was demonstrated by no less an authority than the minister at the Congregational church where I was a deaconess.

It was a rather momentous visit. Not only was Graham Ward my minister; we were friends, and I had been especially close with his wife, who had died tragically of cancer. But eventually, Graham found a new lady love. One day he brought his bride-to-be, Kathy, with two of her small children, over to meet our charming pig, our Frisbee-playing border collie, and our affectionate flock of chickens. It was meant to delight both the lady and her children, rather like a visit to a petting zoo. But it didn't turn out that way.

The pig, then about five hundred pounds, bucked and thundered out of his pen, full of porcine exuberance. Kathy was visibly alarmed. "I didn't know he would be that big!" she exclaimed. Christopher ran to his main outdoor feeding area, and we hooked him up to a tether so the kids could watch him eat a bucket of slops. (When it came to eating, he was a performance artist.) His large tusks did not escape the young mother's notice. She took her daughter's hand. Though her son was fascinated, I was afraid the little girl might cry.

Now, as if to our rescue, the Ladies rushed over to share in the food bounty. "Here's someone more your size!" Graham said to the little girl.

"Watch—you can pat her." As he had done many times with our hens, Graham, dressed in shorts, reached down to pet one of them.

Just like when I stroke the Ladies, Graham started to run his hand down her back. And just as the hens always do with me, she assumed her distinctive squatting posture. This is a well-known chicken behavior usually directed at a member of her own species. It is actually known as a "sex crouch." It's a position that a chicken normally uses to make it easy for a rooster to mount her.

Just then, from another part of the yard, Alex the rooster looked up. The scene looked innocent enough to human eyes: two small kids, three adults, a pig, several chickens, all standing around together on a sunny summer day. But through his rooster eyes, Alex saw a moral travesty, an insult to his roosterhood: the minister was trying to have sex with his hen.

I only had time to cry "Graham, watch ou—" before Alex hit. Spurs first, the enraged rooster flew at the back of Graham's bare calves with the full force of his fury. He broke the skin. His fiancée was of course aghast; the little girl began to cry. Graham tried to assure the children the rooster wasn't mean; he was just protecting his flock. But the kids were having none of it. This was nature red in tooth and claw. I carried Alex ignominiously away to the coop, while Graham, bloodied, ushered his new family to the car. I tried to end things on a good note by handing the shocked children each a fresh egg to take home. Unfortunately, I later learned that one broke on the way back, staining the car's upholstery.

None of our roosters stayed for long. The gentlemanly, long-tailed Lakenvelders, always pictures of vigorous health, dropped dead from their perches within days of each other before their second birthday. Gretchen told us this is not uncommon. Roosters' constitutions are quite different from those of hens, right down to a marked difference in respiration: a hen breathes thirty to thirty-five times a minute; a rooster only eighteen to twenty times. Apparently their different physiologies make the males more likely to drop dead, a phenomenon we named Sudden Rooster Death Syndrome.

The other roosters, we had to ship off. A farmer a few towns over was glad to take them and assured us they weren't destined for the pot. "He's a good 'un!" he said, holding the rooster upside down by his feet. The farmer raised exotic chickens and needed roosters for his many hens.

The Ladies, frankly, seemed somewhat relieved. Though they had appreciated their roosters' food calls alerting them to particularly juicy worms or hidden treasure in the compost pile, there was a cost: the regular annoyance of someone jumping on your clean back with his dirty, scaly feet and biting your comb with his beak—whether you felt like it or not. Sometimes our hens would squawk in annoyance.

Free of roosters, the flock seemed more composed, cohesive, and affectionate. The peace of the henhouse was restored—a feminist utopia.

But even our gentle Ladies have a strange and sometimes disturbing side to their souls. A Speckled Sussex whom the girls named Pickles—the only hen who enjoyed this particular food—revealed to me that the Chicken Universe, though in many ways the sweet soul of domesticity, is inhabited by aliens.

Pickles was a special needs chicken. The moment she arrived as a fuzzy chick we saw she had a bump at the top of her head the size of a small pimple. We soon discerned, with growing dismay, that she had a tiny hole in her skull and that the pimple was filled with cerebrospinal fluid. But she seemed otherwise healthy and normal.

As Pickles shed her baby down for feathers, we began to realize she was a little slow. She seemed to be the last one to notice a food call, for instance, and when the other chickens would come running, she usually brought up the rear. She didn't react normally to loud noises or quick motions: unwisely, she accompanied Howard when he was sawing wood, her neck, I thought, perilously near the blade. Pickles had a tendency to wander off on her own, though not in the spirit of adventure. She always seemed a little lost.

At first, we didn't realize that Pickles couldn't peck straight. When we put down feed, there was usually enough of it that if a hen pecked anywhere within a half-foot radius, she'd hit something to eat. Only later, when we would offer Pickles a small, single morsel from our hand—a tiny ball of pie dough, a piece of carrot—did we perceive her problem. She'd consistently peck about two inches to the right of it, entirely missing the treat.

But no one in the flock ever minded that Pickles was a challenged chicken. We suspected she was the lowest in the pecking order, but nobody had to peck her to drive the point home. Pickles accepted her station without question; she had no ambition to rise in the hierarchy, so there was no reason for quarrel. If anything, Pickles's problems made Kate love her all the more. She'd carry the brown hen around with her everywhere, sometimes tucking her beneath her sweatshirt or sweater, and Pickles would remain perfectly calm and contented there.

One day Howard and I heard a chicken commotion in the yard and rushed out to see. Somehow, on her wanderings, Pickles had sustained a hideous injury. A long slash stretching perhaps four inches along the length of her neck—had she run into barbed wire?—had nearly severed her head. But what horrified us even more was the way the rest of the flock greeted their wounded comrade. Our Ladies, normally such paragons of calm and homey affection, were biting cruelly at the bloody wound, scolding and chasing her savagely.

What was happening? Why would the Ladies do such a thing?

"It's just something chickens do," Gretchen later told me, almost apologetically. "It seems so mean. But they can't help it." That chickens will attack and even kill a wounded flock-mate is so well known that products have been developed to cope with it. One is called Stop Peck. It comes in a bottle like Elmer's glue, looks like clotting blood, and to chickens, apparently, tastes awful.

But I didn't know that then. All I knew was Pickles needed help fast. I scooped her up and held her in a box on my lap as Howard drove us to the vet.

Our wonderful vet, Chuck DeVinne, cleaned and disinfected the wound and gave me some antibiotics to mix in her water. She'd have to be separated from the flock, he told me. Until the wound healed, it would be a target impossible to resist. And in fact, the most immediate threat to her recovery was this same, deep, ancient chicken instinct: to peck at the sight of blood is a drive so strong that a chicken will often kill herself doing so to her own wound.

Chuck fashioned her an Elizabethan collar from a discarded X-ray, wished us luck, and sent us home.

* * *

"There is a deep vein of locked-in behavior in all birds," says Gary Galbreath, an evolutionary biologist at Chicago's Field Museum. Birds are thinking, feeling creatures, but some of what they do is beyond their conscious control, irresistibly carved into their genes.

Lorenz's fellow Nobel laureate, Niko Tinbergen, showed this with a series of striking experiments with herring gull chicks. Normally wild chicks peck at the red spot on the parent's bill to induce the parent to regurgitate food. But they will also, he found, peck at a red dot on a fake "bill" attached to a disembodied head made of cardboard. Tinbergen and his students presented the chicks with a series of options. They found chicks will peck at the red dot even if the head is painted blue. They'll peck at the dot even if the head is weirdly misshapen. And they will preferentially peck at a giant red dot instead of the bill of a real parent if given a chance—even when pecking the giant red dot never produces any food for them at all. For the chick, the red dot acts almost like a push button does for a machine, "releasing" an inborn, preprogrammed pecking response.

In another experiment, Tinbergen showed that a mother oystercatcher, a shorebird, may make an equally strange choice. Normally she lays a clutch of three, but she may lay up to five. But once she begins to lay, if presented with a giant fake egg—painted in natural colors—she will abandon her own. She'll rush to the giant egg, frantically trying to sit on it, every time. She cannot resist.

Konrad Lorenz even observed such mindless behavior in his tame jackdaws, crow-like birds who sometimes demonstrated their affection for him by bringing worms to his mouth (when he refused to eat them, the birds would resolutely stuff them in his ears). But if he took any one of his flock into his hand, the others would attack him. One day, quite by accident, Lorenz made a profound discovery. Coming back from a swim, he was standing on the roof among his jackdaws when he remembered his black bathing suit was in his pocket. The minute he took out the limp black clothing, his jackdaws suddenly flew into a panic, shrieking alarm calls and attacking him with their feet and bills.

It was not that these birds mistook the swimsuit for a fellow jackdaw.

Birds' eyesight is excellent, and they were at close range. There was no mis-understanding. There was no understanding at all. The sight of something limp and black triggered the behavior. The response was as involuntary as a knee jerk.

"Such 'mistakes' . . . have always been emphasized," wrote Tinbergen in his landmark *The Herring Gull's World*, " . . . and it was usually pointed out that 'instinct' was 'blind' and rigid, and made the animal behave very stu-pidly when confronted with unnatural situations." But this interpretation, he warned, is "incorrect and inadequate." Herring gulls are excellent learn-ers; the intelligence of jackdaws is well documented. Yet sometimes, smart birds, capable of reasoning and forethought, are governed by an ancient, genetically determined program beyond their conscious control. Bird behavior is the product of both.

Sometimes birds' actions might look stupid or cruel to us. Even Konrad Lorenz was appalled when, hoping to breed a cross between a turtledove and a ring-necked dove, he put these two symbols of peace together in a spacious cage, only to find two days later that the female had flayed the male within an inch of his life. (Of course, had they not been caged, the victim could have escaped relatively unscathed.)

But we err to judge on the basis of an instance or two; the bird lineage has made its choice over millions of years. Over that vast span of unforgiving time, what might seem in an experiment like a silly mistake has proved to be, instead, a course of action so right, it means the difference between life and death. At times, it is better to let one's ancestors make the decisions—and the ancestors have left their instructions in the genes. It's a deep wisdom of a sort humans seldom recognize.

Why should chickens attack at the sight of blood? I haven't found a fully satisfactory answer. Possibly because the sight of blood usually signals meat, a rare treat, which should be quickly eaten. Perhaps because a severely wounded flock-mate might attract predators—better to drive away one member than endanger the whole group. I don't know. But although dismayed by my Ladies' inexplicable savagery, I was not angry. I was awed—reminded how privileged I am to be allowed to travel in this alien, avian universe.

* * *

Pickles made a quick recovery, exiled to the downstairs bathroom, where she perched on the rim of the sink—a situation that alarmed dinner guests who did not expect to find a chicken in an Elizabethan X-ray collar watching them when they used the toilet. Fully healed, she was accepted back into the flock without a problem. Life with the Ladies resumed its peaceful, cheerful course.

The flock gained new members as we tried new breeds: Dominiques, the black-and-white ancestors of the Barred Rocks, and Black Australorps, a cold-tolerant Australian breed quite similar to our original Sex Links. We lost some old birds to age and to visiting predators: a mink, an ermine, a hawk.

Equally sad for us, the little girls and their mother moved away, to be nearer their grandmother in Connecticut. Though the house next door sat empty, our hens continued to consider that yard theirs—but not the yard across the street. Crossing the street may never have entered the minds of the new chicks; if it did, perhaps the older birds dissuaded them. The boundaries of their territory seemed to be part of the flock's cultural memory.

New neighbors moved into the empty house. At first we were nervous— what if they had dogs or kids who would chase our Ladies or harass our pig? What if they didn't like our compost pile, which was convenient to our barn but only yards from the boundary of their property?

But we needn't have worried, for they turned out to be Bobbie and Jarvis.

In their previous home in upstate New York, the couple had raised three sons, numerous pigs, and several flocks of chickens. Bobbie told me her chickens' individuality and intelligence had taken her by surprise. In the summer, the hens liked to hang out beneath the kitchen window, and Bobbie was amazed to discover why. She used to keep a classical station on the radio when she worked in the kitchen, which they could hear through the open window. Apparently, they loved music. When the radio was not on, they preferred to forage near the picnic table. All except Leticia, a pretty White Rock, whom Bobbie had named after her great-great-great-grandmother. Leticia held herself aloof from her flock-mates, choosing instead to forage alone—unless she could be with her human family.

The Coffins' hens had been tended by a series of handsome and coura-geous roosters, whose photos they still keep. The last of them had given his life for his flock, trying, unsuccessfully in the end, to defend them from a marauding dog. Shortly after the tragedy, Jarvis retired from his job brokering waste fibers to paper companies, and the couple moved here.

Immediately, Bobbie and Jarvis volunteered to help look after our pig and chickens. Beautiful, slender Bobbie loved to bring the hens treats. Soon she was buying cracked corn at the Agway just to feed our hens. The couple visited our pig and chickens so often that Jarvis, handy and enormously strong, built a wooden walkway over the boggy area between their yard and ours.

One day, we returned from a three-day trip to discover Jarvis had replaced the rabbit hutch we'd used as a communal nest box with individual nest boxes covered with a long, sloped, hinged lid. Another time he rebuilt their perches, offering varied shapes and thicknesses for greater choice and comfort.

None of this kindness was lost on the Ladies, of course. The area immediately surrounding the barnyard provided the hens with an excellent vantage point from which to observe their benefactors. (Our barn is in fact closer to the Coffins' house than to ours.) The Ladies soon figured out where Bobbie and Jarvis kept the cracked corn. They listened for the distinctive slam of the Coffins' porch door. They watched for any sign that either of them—or any of their visiting grandchildren—might be coming over. If the Ladies were occupied elsewhere and failed to notice, Bobbie would call them, and they'd come running over. Jarvis would call individuals by name. After the Cabot girls left, he had taken over the chicken-naming responsibility: he named one older Dominique "Mother Hubbard" because she liked to look after new chicks and named an Australorp he considered particularly bossy "Hillary," because Jarvis is an ardent Republican.

From their second-storey back porch, the Coffins had an excellent view of our barnyard. But when we were away and the hens were under their sole watch, that wasn't good enough for Bobbie. To guard against predator attack, Bobbie used the same device she used to surveil sleeping grandbabies when they came to visit: the baby monitor. Any commotion would alert her to trouble and she could come running. Howard and I thought this was a great idea, and we bought one too. Not only did it ensure the flock's safety; as I wrote my books, my words and thoughts would be bathed in the soothing sounds of the calm chicken voices in our barnyard.

But those days would soon give way to a new era. Our lives would change radically with the arrival of more new neighbors: the Chicken Whisperer and her Rangers.

Elizabeth Kenney came recommended by a close friend and fellow animal lover. We needed a tenant for the half of our house we rent out as an apartment; Elizabeth was getting a divorce and needed a new place. But not just any place would do: it had to have a barn.

Elizabeth, thirty-six, worked with our friend as a record keeper for a local home care and hospice organization for the elderly. When I met her, I could see she must be very good at this: she's smart and empathetic, with an air of calm that instantly puts you at ease. But her job is not her passion. Her passion is her chickens.

After we first met, Elizabeth sent me their photos. The flock of twelve included Barred Rocks, Rhode Island Reds, and Black and Red Sex Links. In the divorce, she got custody of the flock, which she called the Rangers.

There was plenty of room for them in the barn. Two summers before, our pig, fourteen years old, had died of old age in his sleep; Elizabeth could move the Rangers into his old stall. Her dedication to the flock was evident the day she arrived that October. A friend came to help her erect a prefab chicken house within Chris's old pen. The guy had once been part of a carpentry team featured on the TV show *This Old House,* and the coop featured pane glass windows, front steps, a linoleum floor, and a ramp leading to two tiers of nest boxes, curtained for maximum laying privacy. The perches were covered with soft towels that Elizabeth laundered every week. The walls of the coop were adorned with artwork—pictures of chickens, of course.

Elizabeth's side of the house was soon decorated in similar fashion. The upstairs windows were curtained with fabric printed with hens. Visitors were welcomed with a chicken doormat. Chicken photos adorned the walls, including a recent Christmas card for which she had won an award. It was a chicken dinner—hens feeding happily from porcelain plates atop a candlelit, tableclothed dinner table. When I came over for my first long visit, she pulled out a thick album, labeled "The Gang: May 05" so she could show me her first flock's baby pictures. (More chicks came the next year, who were similarly documented, to make up her current flock of twelve.) "Here's their first dust bath!" Elizabeth said. Like many birds, chickens love to roll, kick, and fluff their feathers in powdery dust, which must feel as good to them as talcum powder after a bath. Turning a page: "Here's Peanut, that's Averil, there's Jan . . . I just love Jan's face . . ."

Every one of Elizabeth's chickens is named. To her, the identity of each is as obvious as a person's. For instance, "My Jan is nothing like Goldie," she told me. "Goldie's sweet as pie, hyper and curious, and maybe not too bright. Jan's so smart. Jan will stand and sometimes she'll stick her head in my pocket to get the bread out. She knows where it is. She's a wise old lady— an old lady in a young girl's body. She's number one in the pecking order, but she's not aggressive up there. It's like she's got her silent power and only asserts it when somebody really gets out of line, and usually it's just Nikki. And Nikki's second in line . . ."

Elizabeth spends at least an hour a day playing with her hens on weekdays, and spends much of the weekend with them. She knows each chicken's distinctive voice. "I can tell, if I'm turned away from them, who's saying what," she told me matter-of-factly, "and a lot of times what they're doing when they're saying it. If they are making curiosity sounds, I'll know whose voice it is, and if she's happy or not. I can tell you Peanut is going to walk over toward the corncob, and Nikki's going to give her a nasty growl, and then Peanut's going to run that way . . ."

Shortly, I would see that this was true, and more. Some of her chickens, I would discover, say certain things in a special voice, and use it only for Elizabeth. Others perform tricks on her command. I began to realize with growing amazement whom Fate had brought to us as a tenant: we began to call Elizabeth "the Chicken Whisperer."

It's a warm, overcast autumn day, the close of an Indian summer. Elizabeth has just gotten home from work, and the Rangers are out. Most of the day they spend outside, behind the turkey-wire fence she has erected by their coop. In the afternoon, they listen for Elizabeth's car, whose sound they recognize, and know she'll let them out. Now, says Elizabeth, "they're ready to rumble!"

Elizabeth Kenney and the Rangers

Elizabeth issues her "Come to me" call—it sounds like a slower version of the tufted titmouse's whistled "chiva-chiva-chiva"—and they come running, wings out and beating to propel them forward more quickly.

"C'mon—charge! Go get 'em!" Elizabeth, as usual, has brought treats. She sits down on the lawn and holds out an apple. Wide-eyed Goldie, second lowest in the pecking order, arrives first, hops on Elizabeth's black pants, and starts stabbing the apple with her beak. Next Rudy, the red Sex Link who often escapes from the pen during the day, and after her top-ranking, smart Jan, a Rhode Island Red. Jan draws herself up tall above Rudy, then looks down her beak and pecks her meaningfully before advancing upon the apple. Now Lola, a Black Sex Link, comes running, and next Peanut, a Barred Rock.

Peanut is possibly Elizabeth's favorite chicken, if she can be said to have a favorite. When Peanut was a year and a half old, she developed a blockage in her crop—the muscular compartment where birds store and soften their food before it moves on to the gizzard, the main organ of digestion. When medicines failed, Elizabeth begged her vet to operate on the chicken. The first surgery failed. The second was a success.

During her long recuperation, Peanut lived in a spare room in the house, and Elizabeth fed her twice a day in the kitchen with a syringe. "Every morning, when I'd get breakfast, she'd hear me and purr like a cat," Elizabeth told me. "She'd fluff up and greet me and be very happy." Every night, as Elizabeth stroked her, Peanut would sit in her lap. When she bent over the hen, wisps of Elizabeth's shoulder-length auburn hair would dangle down. Peanut would take the errant strands in her beak and gently tuck them behind Elizabeth's ear.

The special closeness the two shared as Peanut recuperated still persists. Elizabeth bends down to kiss Peanut on the head. In response Peanut smacks her beak, making a kissing sound. "She only does this with me, only when I kiss her," Elizabeth explains. Lacking lips, Peanut is doing her best to kiss back.

Several other hens say things in a special voice, only to Elizabeth, never to other hens. "C'mon, Janny, c'mon, boit-boit!" she calls. Jan hops onto Elizabeth's outstretched legs. "Boit-boit-boit!" Elizabeth whispers into her ear. Jan cocks her head as if thinking for a moment. "Boit-boit-boit," the hen replies softly.

Another hen, Elizabeth tells me, says "duff-duff-duff." She'll show me later—but first she wants to demonstrate Jan's favorite game, which the hen herself made up. "Watch this," Elizabeth says to me, and then turns to Jan: "Too-too-too-too-too-too!" Jan seems to know what this means; she looks at Elizabeth expectantly, first with one eye, then the other. Elizabeth drops a small piece of bread from her pocket in front of the red hen. Jan picks it up in her beak. But she doesn't eat it. Instead, she ostentatiously drops it, then raises her head to broadcast loud clucks. "She's calling the others, just like a rooster would," Elizabeth explains. Hens look up from all around and start running. Goldie, still close by, is the first to arrive. She spots the treat—but before she can reach for it, Jan pecks it up and swallows it. That's the game. "She thinks this is really funny!" says Elizabeth.

The hens have other games as well. Elizabeth has brought a newspaper to demonstrate. "Whenever I try to read the newspaper, they come out and they MUST destroy it!" She opens the *Monadnock Shopper News* on the ground before her and, pretending to read it, mutters, sotto voce, "Gee, I'd like to read this article . . . I really need that phone number there in that ad—I'll have to write it down. I sure hope nothing happens to the paper!"

A handful of hens immediately rush over. Elizabeth calls the rest. Rudy steps on the paper. Nikki, a Rhode Island Red, starts shredding the paper with her feet, holding it in her beak. "C'mere, Soot!" calls Elizabeth to a Black Sex Link with a tall comb. The bird looks up, comes directly to Elizabeth, and starts pecking at the print. "Get that paper! Get that paper!" Elizabeth says encouragingly.

All the hens were raised on newspaper bedding, she explains, and they still love to scratch and shred it as they did as chicks. "Oh, look, an ad for Ocean State Job Lot," she says to the assembled group. "Sometimes your roost towels come from there."

"They like the Job Lot advertising section," she tells me, "because it's so colorful, I think." Goldie now arrives and starts shredding the paper with her feet. Peanut comes and pokes a hole in it with her beak.

Peanut, Jan, Nikki, and Soot all know their names. "The commands they know are impressive," says Elizabeth. "When I was carrying wood shavings for the floor of the coop and they would mob me, I would say to them, 'Back-back-back'—now all I have to do is say that and they stand out of the way." They will even back off, when she asks them, when they are massing at the opening to the pen or when Elizabeth is coming into their henhouse. They also respond to "in-in-in" and will rush back to the coop if they're near it. To call them from farther distances Elizabeth gives the chiva-chiva-chiva whistle.

"If I'm sitting, I'll say 'Coming up' and they'll jump into my lap or my back. They also know the show of empty hands and 'all gone' and will walk away, realizing there is no more bread or other treats," she tells me.

Some people often attribute to their pets (and their children) intelligence and insight they don't have. Animal people are often accused of anthropomorphism, projecting human motives and emotions onto animals. But this is not the case with Elizabeth. She does not think of her Rangers as feathered people. She understands that they are different from us; they experience the world in ways we cannot imagine. They can see polarized light; they may hear in a different range; they can fly. They think and feel—but not always as we do. Right now, for instance, several Rangers, including Jan, are molting their feathers. Unlike my Ladies, whose molts have always been subtle affairs, molting Rangers sport ugly red bald patches and itchy-looking areas where new pinfeathers are coming in. When one molts, often the others will try to exclude her from the coop. Elizabeth has to intervene.

"There is lack of pity in all birds," Elizabeth says thoughtfully, as she feeds Peanut a bit of bread. "Chickens have no sympathy. People can't relate to it, so they don't like it. It's something we're not used to. But I like that about them," she says. "There's something very brave about them."

I'm deeply moved by Elizabeth's relationships with her chickens; I'm impressed by all the words, tricks, and games they know; I admire her astute observations. My affection for her and her Rangers grows with each day.

There's only one problem: our two flocks hate each other.

I didn't initially worry about how they would get along. After all, Elizabeth's Rangers were fenced, and thus, except for Rudy's daily escapes, confined by day to the roughly ten-foot-by-twelve-foot chicken yard she had erected outside their coop. My Ladies were never fenced, free to range freely over the property as they chose. Only during the hour or so Elizabeth spent with her hens in the afternoons, and during their time together on weekends, would the Rangers and the Ladies be loose at the same time. There would be little opportunity for the two flocks to fight. Because my Ladies had always accepted new chicks added to their flock with alacrity, I even hoped they might welcome the Rangers.

But I was wrong.

No fights ever broke out. No feathers flew. But after the Rangers moved in, the Ladies, to my great surprise, simply abandoned most of the territory the flock had held for twenty years.

They moved into Bobbie and Jarvis's yard.

I asked the Chicken Whisperer what she thought was happening. She admitted her Rangers were the aggressors. They were taunting my Ladies from behind the fence!

"Through the fence, they act like roosters, with their wings down to the ground like 'OHHHHH—I'm going to kick your tail!' They love doing that. They're really happy doing that. They'll leave their food to do that.

"I think chickens enjoy being in control and dominating others," she continued. Her beloved Peanut, the lowest in the pecking order since her surgery, is one of the worst aggressors: when one of my Ladies comes to investigate, Peanut raises her hackles and leaps to spar at the fence between them.

Now the Ladies never venture past the Rangers' fence. They visit the compost pile when the Rangers are enclosed. But when the Rangers are out, the Ladies take refuge next door.

I viewed the situation as a clash of cultures. Could this be possible? I asked Elizabeth.

"Every flock is quite different," Elizabeth agreed. She found this when she raised a second batch of chicks who came to complete the current flock. The older birds were far more domineering, exploratory, and brave. The difference between flocks was even more obvious comparing the Rangers with the Ladies.

The Rangers peck one another frequently and sometimes even peck Elizabeth—one pecked her by her eye, bruising and breaking the skin. Mine peck my palm when food is on it, but in my two decades of raising chickens, none has ever pecked my face. And though Elizabeth's hens are very affectionate, they do not like being stroked like mine do; unlike the Ladies, the Rangers never squat before her asking for a caress. Instead, they stand still before her, waiting to be picked up and kissed.

My Ladies have always been extremely calm and peaceful. The Rangers are more vocal, more aggressive, more domineering. The Rangers are rabble-rousers, always ready to rumble. My father-in-law would have said they were always making a *tzimmes*.

Everything the Rangers do is writ large. My hens are gentle, subtle; they are Ladies. The Rangers are drama queens.

Does this sound like anthropomorphism? Am I projecting onto chickens traits that belong to humans alone? How can it be that birds—a lineage that separated from that of the mammals more than 300 million years ago—are as individual as people? How can these birds share with us intelligence, reasoning, foresight, memory—and in other ways slavishly obey blind instinct? Can creatures more closely related to lizards and crocodiles than to people actually have culture?

To me, it's clear: though none of my original Black Sex Links of twenty years ago survives, their calm culture has persisted far longer than any individual's lifetime. And it has persisted through generations of unrelated chickens of different breeds.

My travels in the Chicken Universe have been a portal to an unknown kingdom. All of us see birds every day, and chickens are among the commonest birds we know. Yet again and again I am reminded how movingly like us birds can be—and how thrillingly different.

The longer I watch them, the more clearly I see how rich and varied their lives are, as fraught and joyous and changeable as our own. Last spring, to my dismay, I saw something I had never before witnessed.

My hens were crossing the street.

Cassowary

Birds Are Dinosaurs

Unfurling buds coiled like fiddleheads, monstrous ferns rule the forest, as they have done for 320 million years. Some ferns grow trunks fat as trees; others fling their fronds skyward for twenty feet directly from the ground. Tasseled club mosses hang from branches and rocks; vines drape from branches. The Wet Tropics of Far North Queensland, Australia, may be the oldest rain forest in the world—older than the Congo, older than the Amazon—and harbor strange and wondrous lives. Lumholtz's tree kangaroos—true kangaroos who climb into trees, leap thirty feet to the ground, and bounce away—live here, as do squat, two-foot-tall wallabies called pademelons, who signal one another by thumping the ground. This, too, is home to the red-bellied black snake, who grows up to six feet long and is poisonous but, my companion field biologist David Westcott assures me, is "really sweet. People do get bitten, but these snakes are so tolerant, they wait until the last second to bite you. From now till the last six months of the year, you'll have your foot thirty centimeters away from one every day."

In fact we had seen one crossing the winding road leading to David's study area, its onyx scales glistening in the rain.

I was grateful for the sight of it, but I hadn't come to Eden to look for the snake. I had come for something even more prehistoric: a six-foot-tall, 150-pound creature capable of easily slashing to death any enemy with a five-inch dagger-like claw on the inside of each three-toed foot. Standing on tall, scale-covered legs, its blue head topped with a tall helmet of bone, it's instantly recognizable as a dinosaur. Scientifically it is known as *Casuarius casuarius,* a Latinized version of one of its Papuan names, *"kasu weri,"* meaning "horned head." The common name is southern cassowary. Because its big torso is covered in black bristles, hiding its stunted, stumpy forelimbs, you might never guess it is a bird.

"Keep your eyes peeled," David had told me as we began the coiling drive up through the rain to the park. "It would not be impossible for one to run across the road right in front of us. They like roads." But in the forest, even if we don't glimpse the cassowary itself, with luck we might see their scat or their nine-inch-long, three-toed tracks. Their footprints look exactly like the fossilized trackways of young *Tyrannosaurus rex.*

David knows this forest intimately. He has been studying the cassowaries here, in this six-square-kilometer tract of Wooroonooran National Park, for twelve years. Yet David is the first person to say that little is known about his study subject. "Everything is difficult about them," he told me. "They're hard to find, hard to follow." The one woman he knows who set out to do a Ph.D. on them was able to get good data on only five cassowaries—so she wisely ended up doing her dissertation on the interactions between cassowaries and people instead.

Cassowaries are difficult to radio collar. Aiming for the big muscle of the drumstick, a researcher shooting a dart gun uses medetomidine, a drug used to tranquilize rhinos and elephants. Yet even a dose hundreds of times more than would kill a human won't knock out a cassowary. Once the drug takes effect, David said, the cassowary "just stands there. And if you touch it, it'll start up again." He found it was easier just to jump on top of the cassowary and wrestle it to the ground. And how easy was that? "I didn't have any grey hair before I had to do that," the fit, forty-six-year-old ornithologist said, "and I haven't gotten any more since I stopped."

The radio telemetry study ended in 2000. "It was enormously expensive and failed so often," David explains. The transmitters, glued to cattle ear tags attached to the loose blue skin on the back of the cassowary's neck, dropped off or failed, or the cassowary disappeared. Now, he says, he attaches the devices occasionally to birds who have been injured, often by cars or dogs, nursed back to health, and released. (Alas, cassowaries in rehab facilities only survive their injuries about a third of the time.) Scientists have learned disappointingly little about these mysterious creatures.

I'd have to be extremely lucky to see one on my first day out. In fact, on this rainy day, it seems I might be extremely lucky to see *anything:* when David pulls over the Land Cruiser to a scenic overlook, we see nothing: not the Pacific Ocean, not the Atherton Tablelands with their rich farms on volcanic soils—only the mist and the gentle, soaking, steady rain. "Liquid sunshine," David calls it. He's the cheerful sort.

As soon as we start down the slippery Bartle Frere Walking Track, my glasses steam up. But I dare not take them off. I need them for protection from the flora, along with a long-sleeved shirt and long pants tucked into my boots, no matter how hot it is. Having worked in the Australian scrub years earlier studying another flightless bird, the emu, I'm well aware of some of the unpleasantries of a continent famed for its natural poisons. As we start our hike, I ask David about toxic plants. Are there any I should watch out for? Oh, yes, says David; there's the stinging tree. "There are trees that sting you—four species in this forest," he tells me matter-of-factly. The trunk, petiole, and heart-shaped leaves are covered with hollow silica hairs, each containing a gland full of neurotoxic poison so painful that, he relates, "it's completely incapacitating." Searing pain can persist for twelve to forty-eight hours; after that, the silica hairs continue to burn for months, every time there is a change in temperature. Then there are the plants with spines. The most famous is the aptly named wait-a-while, a climbing palm whose thorny tendrils impede your progress, unless you are willing to let it tear your clothes and flesh. "And if you get one in your eye," David adds, "you won't be seeing much." He doesn't mention the other thorny species, at least four of which I think may have nailed me through the cloth of my pants. But later, I discover that most of the pain I had credited to them was probably due, instead, to leech bites. Most leeches have a natural anesthetic in their drool, but these, David explained, are "a bit amateurish in terms of anesthetic. They seem to bite, and you notice that they're there."

I do notice, but there is nothing to be done about it. At any given moment, I can see more than a dozen half-inch, thread-thin leeches crawling on my pants, but to stop to remove them would be pointless. Another dozen would inch onto me in the interim. And I haven't come halfway around the world to patrol my pants for parasites. I am desperate to see a wild cassowary.

Finding emus had been much easier. My study of them, during the Australian winter of 1984, began when three subadults walked up to me as I was sampling plants in the desert scrub of South Australia. Five feet tall and weighing about seventy-five pounds, their long, brown, twin-shafted feathers looking rather like shaggy fur, they approached me within twenty-five yards, carelessly picking at grass with their goose-like beaks, and then calmly strolled out of sight. I'd never seen wild birds so huge, so powerful, so outlandish, so near—it was overwhelming.

I had to find out everything I could about them. At first I never dreamed I would see them again. I thought the best I could do was study their scat—at least I had a good chance of finding that, as it doesn't move around, and it would at least tell me what they had eaten. But I soon discovered that in the outback, with its vast, flat expanses, I could easily find the birds themselves, too. I spent three blissful months sleeping in a tent at night and following the three subadult emus by day, almost every day, recording their every movement.

Here in the rain forest, though, it seems we can't even find a cassowary's footprints or scat. Instead, we find the two-toed tracks of feral pigs—wonderful animals like my Christopher, social and smart. But they don't belong here, and they destroy the fragile soils of the rain forest, compete with native creatures for food, and destroy cassowary nests and eggs. Their introduction is just one of the insults this land has suffered. Though the Tablelands were first settled by miners looking for gold in 1875, the rich soil went largely untilled until after World War II. During the war, the area had been a jungle-warfare training area for Allied troops, who left their tanks behind. Farmers found the machines could be converted into tractors, and large-scale clearing began.

Thanks to the farmers and their imported plants, the remaining forest is now beset with some five hundred species of introduced weeds. They've flourished in the mayhem created by Cyclone Larry in 2006. Before that storm flattened a trail 150 kilometers wide through World Heritage rain forest along Queensland's eastern coast, David tells me, the canopy here was thirty meters high and completely closed. Had I come two years earlier, the green ceiling above us would have blocked out the sun, if there was any, and shielded us from the rain.

Even in its diminished state, the forest retains at least some of its primeval splendor. Some four thousand plants have been cataloged in these rain forests, including eighteen hundred species of trees bearing the fleshy-fruited seeds that cassowaries love. A single half-acre plot might have up to two hundred different species of trees. It's perfect for cassowaries. David has no idea how many live here. Possibly dozens.

But where are they? For three hours we hike, slipping through the greasy mud, tangling our legs in vines, ripping our clothing on thorns, but all we see is how easy it is for a huge black creature to go undetected in a dark forest. If a cassowary is anywhere near us, the hissing rain hides the sound of its movements, dissolves its footprints. Only when we have nearly completed our loop and are just yards from the lot where we parked do we notice the seven-inch-diameter pancake-like dropping, littered with half-inch oblong red seeds. David knows right away what the cassowary had eaten: the fruits of *Aceratium dogrelli*. The closest tree of that species is about a two-hour hike away, down *that* path, he tells me. He gathers the scat in a ziplock bag to bring back to the government-run biological research station where he works in Atherton. A lab technician will later separate out the genes in the bird's DNA that distinguish it as an individual; with enough samples, David hopes, he can begin to estimate how many cassowaries actually live here.

David is pleased. As he drives us back to his office, I note with envy that his clean-cut, square-jawed good looks and cropped sandy coif haven't even been rumpled by our hike. My soaked hair plastered to my head, I know I look a mess. But I am unprepared for the look of stricken horror on the faces of friends when they come to pick me up at the research station parking lot. I glance down at my clothing: the lower half of my pants is soaked in blood and a huge red stain is spreading over the front of my shirt. The leeches ignored David, but in the mysterious way that mosquitoes seem particularly attracted to some people, they seem to adore me. I look like I have been shot.

Leech bites bleed for hours because, in addition to anesthetic, their drool contains a natural anticoagulant to extend the feeding opportunity. The thread-thin creatures, bloated with my blood, are now fat as slugs and dropping out of my clothing onto the floor of my friends' car. Back at their cozy home, after I step out of the shower, I find myself standing in a pool of my own blood. But if this is the price I must pay to see a dinosaur in the twenty-first century, I will gratefully pay it.

On the other hand, I could instead just glance out any window. Although the cassowary looks more like a dinosaur than some dinosaurs did in the Jurassic, it's not the only one around. Dinosaurs flock to our bird feeders. People eat them for Thanksgiving. Dinosaurs crap on statues in public parks, sing from cages in people's homes, perch on telephone wires, and crow at dawn on farms. Most paleontologists today agree that the dinosaurs did not go extinct, as we were taught in grade school. Instead, they became the most diverse group of land vertebrates on the earth—the world's ten thousand species of birds.

How relieved I would have been to know this as a child. I was not the least bothered to discover that Santa was a fiction, but finding that the dinosaurs were extinct was a devastating blow. Upon learning the news, I cried in my room for hours; not even our Scottish terrier, Molly, or my parakeet, Jerry, could console me. Like many kids, I adored dinosaurs and had a large collection of plastic models as well as several plush and mechanical toys (one meat eater, named King Zor, even shot suction-tipped darts from its mouth) whom I gathered into a dinosaur village I called Purplenoiseville. These were my favorite playmates before I grew old enough to go to school. I loved imagining the sounds they made, the way they moved, how they interacted with one another.

As it turned out, my childhood fantasies were just slightly less accurate than the theories of the paleontologists at the time (who, I discovered decades later, had put the wrong head, that of *Camarasaurus*, on my beloved

Brontosaurus skeleton at the American Museum of Natural History—and later heretically renamed the Bronto altogether, calling it *Apatosaurus*). In the 1970s and '80s, an abundance of fresh fossil finds—well preserved, beautifully intact, and newly articulated—have given dinosaurs a whole new life. No longer are they all considered dull, lumbering leviathans. Many were fast, agile, and smart.

More surprises came with new fossil finds at the turn of the millennium, especially from China. The timid titmouse and fluffy chickadee, it turns out, are close relatives of the largest, most powerful carnivores that ever lived. Their direct ancestors were bipedal, meat-eating dinosaurs with three toes, air-filled bones, and wishbones like those from the Thanksgiving turkey. And long before the appearance of *Archaeopteryx*—whose scientific name means "first bird"—many of these smart, swift dinosaurs sported feathers, the defining characteristic of birds. Baby tyrannosaurs, it now appears, were covered in down. And although *Jurassic Park* reflected the latest science of its time, had it been produced after the revolutionary fossil finds of 1999, its *Velociraptors'* hands, arms, and tails would have all been festooned with feathers.

Although they already had feathers 170 million years ago, the earliest bird-dinosaurs did not fly. But the dinosaur family to which they belonged, the dromaeosaurs, already possessed the body plan that would make dinosaurs into birds: a relatively large skull ready to accommodate a bigger brain; long arms that, in many species, could be folded, wing-like, against the body; large hands with three long fingers, which would become the tips of wings; a long tail, which could eventually function as an aerodynamic stabilizer and rudder during gliding; and forward-facing eyes for binocular vision and swiveling wrists, both of which would later prove essential for flight.

The ancestry of birds was hotly debated for centuries. The very traits that define birds—feathers, and hollow bones of relatively small size—plagued early paleontologists' pursuit of the mystery, for these structures do not easily fossilize, and when they do, they are seldom intact. Nonetheless, similarities between birds and the giant reptiles that had ruled the earth suggested links to gifted thinkers long before solid evidence was in. As early as 1867, Darwin's champion, Thomas Huxley, presented birds as "glorified reptiles." Today's Dinosaur Institute director Luis Chiappe titles his book on avian evolution *Glorified Dinosaurs.* Texas paleontologist Timothy Rowe and Lowell Dingus, who directed the recent renovations of the fossil halls at the American Museum of Natural History, titled their book on the same subject *The Mistaken Extinction.*

Supplementing new fossil evidence, new data thanks to DNA technology seem to have largely settled the question. In 2005, a North Carolina State University paleontologist, Mary Schweitzer, found soft tissue in the femur of a *Tyrannosaurus rex*—tissue that scientists previously thought would have degraded eons ago. From it, Schweitzer was actually able to sequence DNA from the tyrannosaur's 68-million-year-old collagen proteins. What living creature's DNA did the dinosaur's most resemble? Barnyard poultry just like my Ladies at home (prompting *Discover* magazine, reporting the findings in 2007, to headline the story "Did *T. Rex* Taste Like Chicken?").

Today few serious paleontologists question that birds arose from dinosaurs. Increasingly they agree on an even more surprising conclusion: that birds, rather than meriting a separate class, Aves, in the scientific organization of life, should be classed in Reptilia, within the Dinosauria, as the very successful surviving dromaeosaurs.

In other words: birds are living dinosaurs. To the nimble likes of predatory *Velociraptors,* birds owe their speed and their smarts. To dinosaurs, they owe their otherworldly appeal—and as well, surely, some of their transcendent mystery and beauty. For this is one of the great miracles of birds, greater, perhaps, than that of flight: when the chickens in my barnyard come to my call, or when I look into the sparkling eye of a chickadee, we are communing across a gap of more than 300 million years.

Of all the world's living birds, cassowaries alone share with dromaeosaurs a distinguishing characteristic: a razor-sharp killing claw on each foot. When paleontologist John Ostrom first described the killing claw in a species called *Deinonychus* ("terrible claw") in 1969, he interpreted it as a blade-like slashing weapon; others have since argued it might have been better employed for stabbing. In either case, this feature is part of what made *Velociraptor* such a fearsome predator—and has earned the cassowary the reputation as the world's most dangerous bird. In the Sydney airport, I noticed that *Australia's*

Dangerous Creatures for Dummies devotes 3 of its 324 pages to the southern cassowary, noting it kills more people than any other bird. (The other large ratites, including ostriches and emus, have been known to kill people, as have certain raptors—usually defending their nests.) Adding to its cachet, the cassowary kills in a showy, bloody manner: even at a placid standstill, the cassowary towers over most people, but when sufficiently annoyed, the bird leaps another five feet into the air and then comes down to rake its feet down your body, the dagger claws slicing you open as deeply and effectively as a switchblade.

Friends at home wondered why I wanted to venture into the dark rain forest hoping to encounter a big black creature who might eviscerate me. Actually, I explained, cassowaries don't kill people very often, and when they do, the people generally bring it on themselves. The last time a person was killed by a wild cassowary in Australia was back in 1926. Two teenaged brothers, Phillip and Granville McLean, went out with their dog to retrieve some horses who had wandered into the bush. Unwisely, their dog attacked a cassowary along the way. Fourteen-year-old Granville rushed into the fray and was kicked in the leg; his seventeen-year-old brother Phillip then hit the cassowary with a bridle. The irritated bird kicked him in the neck and severed his jugular vein. Later into my trip, I heard that another Australian teen had been killed more recently by a captive cassowary at a zoo. I couldn't verify the story, but my informant considered the victim a casualty of natural selection. According to his story, the boy was trying to steal the bird's eggs—each a beautiful greenish blue, weighing 1.5 pounds. The would-be thief's eviscerated corpse was found lying outside the cassowary's exhibit in the morning. The cassowary's eggs were safe.

Many more people are killed by cassowaries in New Guinea, which boasts three species—the tall southern cassowary as well as the handsome northern and the fifty-pound dwarf or Bennett's. There are no good statistics on this, and that's not surprising. New Guinea's small villages are so isolated from one another that people speak more than one thousand different languages. There are few roads, and almost no centralized services to connect them. But cassowary attacks are common enough that when I went there, in 2005, even though I spent only two days in villages (and most of the rest in the trackless cloud forest, searching for Matschie's tree kangaroos), I met a family whose matriarch had recently been killed by one. The killings are essentially barnyard accidents. Many villagers capture chicks to raise to edible adults; the meat is so valued that in some villages, a single captive cassowary might be traded for a wife (who is not eaten), or eight pigs (who are). And though the cute, striped chicks are docile and affectionate, once they reach maturity, cassowaries want to be alone. Because they are penned, they can't run away as they would in the wild, so unwelcome visitors may be met with slashing toenails. It's tragic for the people, of course, but who can blame the cassowaries? They are killing people who plan to eat them someday.

In Australia's Queensland, a Parks Service study of 150 cassowary attacks on humans since 1926 found that most of the time, the people were not hurt but merely chased or charged. Cassowaries can and do kill dogs, but this is in self-defense; along with cars, dogs are cassowaries' main predators. Cassowaries occasionally kick doors or windows. They are probably attacking their reflections, like nesting cardinals sometimes do—except in this case the bird weighs 150 pounds and breaks the glass. But in the 150 attacks that Christopher Kofron studied in detail, only five people were hurt badly enough to need hospitalization. This is not to minimize the victims' ordeals: one woman needed nine stitches to close the wound in her forehead and two to suture a puncture in the thigh; a man suffered a broken leg from a kick; and another man needed stitches in the chest, forearm, and scrotum. But injuries are rare and the cassowaries' motives understandable. The study, published in 1999, noted that several of these encounters were cases in which the cassowaries, whom people had previously fed to prevent them from starving during a drought, were probably asking for food. In other cases, the birds may have been defending favored fruiting trees. In ten instances, males were protecting their nests or their babies.

The cassowary father is a single dad. After mating, the female lays three to five green-blue eggs on the forest floor in primitive scrapes of leaves and grass—a nest more like a crocodile's than a bird's—and then leaves the scene, never to return. The father incubates them, alone, with great dedication: for forty to fifty days, he sits on the eggs without leaving them, neither eating nor drinking. After the eggs hatch, for seven months the father, often whistling, leads the striped chicks through the forest, showing them how to find fruits and insects. I would have loved to see the babies, but because I had to time my Southern Hemisphere trip to coincide with the breeding of another bird I was visiting in New Zealand, a giant, flightless, nocturnal parrot, I arrived in Queensland in early April. At this time, males and females begin courting; the previous year's chicks, now tall, brown, awkward "teenagers," are evicted. But there were benefits to arriving in April, when the tropical wet season begins to give way to drier days and the weather begins to cool. It's a time of plenty for cassowaries: figs and pandanus fruit and the fleshy fruits of trees in the laurel and walnut families abound. The giant birds' minds are on romance. The males are free of their kids. I told concerned friends that these factors would surely render any cassowaries I met less irritable.

But in fact, I never worried about getting hurt by a cassowary. What I feared was that I might not see one at all.

8:00 a.m.: I'm freshly showered, wearing every Band-Aid we could find in my hosts' home, as well as all I had brought in my first aid kit. The blood has been laundered from my pants and shirt. I'm sipping coffee in David's white Holden station wagon and excited about the trip. We're on our way to a place with fewer leeches—an area David considers my best shot at seeing a cassowary. We're headed to a resort area just an hour's drive from the famous Great Barrier Reef. Stretching for eighteen kilometers along a palm-fringed coastline, Mission Beach is a nature lovers' paradise, known for its extensive rain forest walking tracks, its large blue butterflies—and its informal claim as the cassowary capital of the world.

"Why's that?" I ask David. "What does that mean?"

"You'll see," David replies mysteriously.

"You mean I'll see a cassowary?" I ask hopefully.

"Well, I can't promise. The problem with cassowaries is," he says, "here's a big black bird that doesn't want to see you and lives in a complex environment where you can't see it."

Yes, I had noticed that yesterday. In fact, I had well understood how difficult my quest might be long before I arrived in Australia, thanks to the scientist who introduced me to David, ornithologist Andy Mack.

Before he became senior scientist at the field station for the Carnegie Museum of Natural History, Andy studied cassowaries' role in the ecosystem of the rain forest of Papua New Guinea. We had a mutual friend. The tree kangaroo researcher I had followed to New Guinea a few years before put me in touch with him. Andy generously invited me to visit him at his new digs at the Powdermill Nature Reserve in Rector, Pennsylvania.

There I found a bespectacled, bearded blond who had just turned fifty but looked thirty-five, with a hilarious sense of humor and a gift for telling stories. Andy even managed to get National Public Radio to feature his study subject, the little-known, fifty-pound Bennett's cassowary, in a short report. "There aren't that many birds out there that can actually, literally kick your ass and kill you," he told reporter John Nielsen on air, "but these ones will."

A cassowary's foot features a dagger-sharp killing claw. The two feet at left belong to a cassowary who was shot by locals; the foot at right belongs to researcher Andy Mack, who was not.

In New Guinea, working at field sites so remote that Andy, and later his wife and collaborator Debra Wright, had to fell trees and build bridges to get to them, he battled tropical diseases ranging from malaria to filariasis, the nematode infection of which the grotesque swellings of elephantiasis is the end-stage sign. He suffered foot rot so bad he couldn't walk for days, and with forceps wrapped in cotton, he regularly pulled leeches out of his assistant's *eyes*. Under these conditions, for twenty years he pursued the secrets of cassowaries. Over the course of two decades, he only caught fleeting glimpses a handful of times.

One of his most dramatic encounters happened the very first week he was in New Guinea. He was scouting for study areas. The first he visited was Varirata National Park, just outside the Papuan capital, Port Moresby. On his first hike in that rain forest, he found himself surrounded. My heart pounded as Andy told me what it was like:

"All around you things are thrashing—they're just bashing through the undergrowth and falling. All these things are waving frantically and getting knocked over. It sounds like a horse stomping. Big animals are chasing each other around. One dashed into sight and out. They're just masters at staying out of sight. Even when you're seeing them, you're not really seeing them. They're black, and it's dark in the rain forest. Something black goes by. Everything's moving around you, and now and then a big black animal appears and disappears quickly, and another appears and disappears, and you couldn't quite figure them out. I think there was a third bird, but it wasn't moving very quickly, because I could now and then get a glimpse of this other one. It was just sort of hanging out. I don't know what was going on there, but it was really exciting . . . It's like the whole forest comes alive around you with these things."

He never saw anything like it again.

Andy employed native people to help him. Local people know cassowaries well, for they hunt them. They kill any adult cassowary they can find—or they hunt down its nests to raid for chicks. They love the meat and use the cassowary killing claw to tip spears and arrows. They use the long, twin-shafted black feathers to decorate headdresses; they sharpen the strong, hollow leg bones to use as knives and spears. So it was no wonder that the cassowaries fled at the sight of Andy's approach. "If they saw you," Andy said, "it was like BANG—away." A cassowary can run thirty miles an hour; it can also swim, crossing rushing rivers with ease, paddling with its strong legs and feet.

"You go weeks and weeks without seeing them," he told me. "If it doesn't move, you could walk within ten meters of a cassowary and never know it's there. You see footprints. You see where they sleep. You find their feathers. And their scat is easy to find." In fact, in cassowary scat—like the pie David and I had gathered at Wooroonooran—Andy found a scientific gold mine. In a series of careful, ingenious experiments, he showed that cassowary scat regenerates the rain forest. And for many dozens of fruiting species, cassowaries are the only creatures in the forest doing so; most of the seeds that pass through their gut are so big no other animal could possibly swallow them. But when the cassowaries do, they transport seeds, in some cases hundreds of meters away from the parent plant, often uphill, where the seeds could never roll on their own, and essentially plant them in a neat packet of fertilizer.

Andy and Debra's twenty-year study of the four hundred species of fruiting plants of the Crater Mountain area of Papua New Guinea still stands as among the most complete ever conducted for any rain forest frugivore anywhere in the world. Andy has probably spent more time in pristine cassowary habitat than any scientist on the planet—but while the cassowaries thrilled and fascinated him, his studies were really more about fruit than the birds who ate them. He would still love to learn more about cassowaries.

"You can track them. I could follow where they were going. But if they hear you, they can just slip away. They are like big cats, stealthily moving in the forest shadows. They're mystery birds, enigmas of the rain forest."

My own prospects of seeing a cassowary were probably completely hopeless in New Guinea, but that's why Andy introduced me to David. For in the very different rain forest where David is now taking me, I'll have an advantage that Andy had lacked: at Mission Beach, instead of native hunters from whom they must constantly flee, cassowaries instead enjoy a dedicated network of fans.

The drive from Atherton to Mission Beach takes only a couple hours, but David has planned some stops along the way. He wants to introduce me to Margaret Thorsborne. "She knows cassowaries," David says, "and who knows? At her house, maybe we'll even see one."

Her "house" is really a series of tin-roofed wooden cottages in the middle of a rain forest. "Fairies Live Here," proclaims a small wooden plaque near one of the doorways. It might be true; there is certainly nothing to keep them out. The doors are all wide open, as are the windows, all of them screenless in order to allow the animals to come in whenever they like.

They do so often. A honeyeater flies in through a window at the same moment we enter the door. Skinks scuttle in and out of the doorway and over the yellow paint of the kitchen floor. Clouds of mosquitoes gather inside, despite the single green mosquito coil burning from a back doorway. In the two kitchen windows are two immense spiderwebs, each with a spider in its center. The larger of the two has legs that span at least twelve inches and a body the size of a small mussel shell. "The dear little thing has had quite a time," our hostess explains to us as she puts a kettle on the gas stove to boil tea. "A bird flew into her web, and yesterday she repaired it, but today it is broken again."

Margaret is a tiny, bird-like lady, at eighty-one slightly stooped but still strikingly beautiful, with thick white hair and wide-set, bright blue eyes radiant with compassion. I notice five tiny orange spiders sharing the web with the huge one. "Are they spiderlings?" I ask.

"No," she answers, "they're males. There were thirteen yesterday, but only five today, I'm afraid."

"Is that because she's eating them?"

"Yes," Margaret says gently. "Yes, she is." Margaret doesn't hold that against her.

Neither did she mind when a blue tree snake entered her home not long ago. "He was lovely," she says, and passes us the photos she took to commemorate his visit. "Here he is, going over the teapot." Another photo shows the snake at home among the many books, most of them about nature, on her shelves. This is a different species from the tree snake whom she found curled up in a folder in her bedroom this morning, she tells us. Her description seems to match that of the Australian brown tree snake that was mistakenly introduced to Guam, where residents are vexed to find the reptiles lunging out of their toilets to bite them and coiled around infants in their cribs. "I was a bit nervous when I first saw him," Margaret says, "but he was very sleepy, and I moved the folder away from the house."

Margaret is a girl of the Limberlost, now in her wise old age. No wonder cassowaries like to visit her here. For the last four years, a male has come to her yard regularly. The bony helmet on his head, called a casque, is distinctively bent and on one side appears rectangular. She calls him Cass.

Cass started visiting one dry season, drawn to the small bowl of water she put out for the other birds. He appeared frantic for water, she said, so she provided a larger bowl: an inflatable plastic kiddy pool, which he seemed to appreciate. The female—much larger than her mate, as is the way of the cassowary—is more cautious. Margaret sees them briefly together, and then both disappear. Two months after her first sighting, Cass arrived in Margaret's yard with his chicks. "He's had four lots of chicks since we met. This year he has three. He's a very good dad," she says, passing us a tray of limp macadamia nut cookies and raisin squares.

Like the cookies, I am sodden from the humidity. It must be 90 percent here, for we are actually in a wetland as well as a rain forest, which also accounts for the clouds of mosquitoes. Margaret's cottage is in the Edmund Kennedy National Park, a 69.5-square-kilometer network of coastal wetlands and rain forests. She and her husband, Arthur, a former teacher who died in 1991, donated the property to the state, with the proviso that Margaret could live there for the rest of her life. She wouldn't want to live anywhere else.

More than forty years ago, she and Arthur initiated annual counts of pied imperial pigeons—finding, when they started, only fifteen hundred—and ensured their legal protection. Now there are thirty thousand. They collected and cataloged more than two hundred botanical specimens from the area. Other scientists have named two species of moss, a vine, a crab, and a spider after them. Through the Thorsborne Trust and her charismatic personality, Margaret continues to champion conservation in the area, especially for cassowaries. "They're such beautiful, stately, regal birds," she says. "When you see them walking quietly, in their natural way, it's a privilege, really. Some people only get to see them run. When they have to run away, it's a shame."

Margaret keeps detailed notes of Cass's comings and goings. She is eager to share them with us. Pulling out her conservation calendar from 2006, she tells us that he had one chick when she first met him on September 11. Then some days later she saw him with two. A couple of weeks later when she saw him next, there were three. David mentions that sometimes—as is the case with emu fathers—cassowary males swap chicks with other fathers, possibly by mistake. But now Margaret is busy looking up this year's records.

"On the twenty-first of February I saw him cross the road. On the twenty-ninth he was on the road to the arboretum. Oh, here! March tenth:

Cass plus three. March eighteenth: they crossed the clearing here and went down the creek. The chicks are getting big, all brown by now . . ."

As she walks us to our car, Margaret points out the "cassowary garden" she's allowed to grow by making sure Cass's seed-laden droppings go undisturbed by the lawn mower. She wants to ensure that Cass, who might live to age forty, as well as his chicks, might always find their favorite food plants nearby. Here, she shows us, is a species of ginger, already waist high; over there, a young fig tree. The ginger has bright orange fruits and blue seeds, she says, and she toddles over to a larger ginger to show us. And now she reaches high, grasping a tall branch laden with ripe fruit—and tenderly bends it so Cass might reach it more easily.

"He's a particularly nice cassowary," Margaret tells us. The delicate octogenarian leaves us with a stunning parting image: "Sometimes I'll be in the yard crouched down weeding," she says, "and he'll be there looking over my shoulder."

Is this the killer dino-bird from *Australia's Dangerous Creatures for Dummies*?

"I get questions from TV crews all the time, desperate to hear how aggressive they are," David tells me as we resume our drive. " *'If you were in the forest and the cassowary charged you, could you outrun it?'* Sure, they're the only bird that kills people, and in New Guinea it happens regularly, but look, they're big. Horses are big. And they kill people, but nobody thinks of horses as aggressive creatures."

"So," I ask David, "what are they *really* like?" It sounds like the sort of question an infatuated fan asks about a favorite rock star, but David's response is doubly important to me. I want to know how wild cassowaries behave for their own sake, but I also want to know for the same reason paleoanthropologist Louis Leakey sent Jane Goodall to Tanzania to study humans' closest living relatives, the chimpanzees. Leakey longed to know what protohumans were like. He had found their bones and stone tools, but, as he always lamented, "behavior doesn't fossilize." Chimps provide a clue. David's understanding of wild cassowaries, I hope, might illuminate an equally seductive mystery: the forever-vanished character of their dinosaur ancestors.

David considers for a moment. "They're timid," he says. "They're sedate. They're interested and curious, though they move through the forest at a sedate rate. They spend a lot of time watching insects. They follow butterflies"—what a lovely thought! *T. rex* engrossed in a butterfly's flight!—"you know, because they eat them.

"They're frustrated carnivores," David continues. "They're irritated that God didn't give them teeth! They'll eat meat at every opportunity. They'll chase down a lizard and stomp it to death and eat it. Those are big whomping feet. Rats, too. One stomp and it's dead. They *love* a dead rat." I had known this, for Andy had explained how he had taken advantage of their taste for rats in his research. To measure how long it took food to pass through the cassowary's gut and how far ingested seeds were carried, Andy used wild-caught rats as bait, placed beneath favored trees. He devised a packet combining a data logger and a radio transmitter and stuffed it into the rat's body cavity (the transmitter's antenna protruding from the anus and tied to the tail). There was no ill effect on the cassowaries or the equipment, for the cassowaries swallowed the rats whole.

Birds who eat mammals whole! Except for raptors like eagles and owls, we seldom think of birds as carnivores. But they are. Robins hunt worms, swifts snatch mosquitoes, kingfishers butcher fish. The warblers and vireos scratching the leaf litter, the nuthatches and creepers patrolling tree bark are all looking for insects to kill and eat. Crows and jays relish meat. Vultures rely on it. Even hummingbirds depend on bugs for half their diet. They take smaller prey, but many birds still follow the dietary traditions of the carnivorous dinosaurs, who as well as eating other dinosaurs also fed upon our ancestors.

To do so, they had to outwit them. The mental demands of hunting prey in a complex environment may have driven the rise of intelligence among the dinosaurs who became birds. Relative to body size, the braincases of the early birds like *Archaeopteryx* were dramatically larger than those of the pterosaurs, the flying reptiles. Both needed good balance and coordination to fly. Sankar Chatterjee, curator of paleontology at Texas Tech University, points out that *Archaeopteryx* had a more demanding, variable life. Unlike the pterosaurs, who mainly hunted fish and nested on cliffs near seacoasts, *Archaeopteryx* inhabited a complex, multilevel forest niche. These early birds needed advanced vision and hearing. Against a confusing background of leaves, branches, and sky, they had to make complex judgments based on sight and sound. As I read Chatterjee's 1997 book *The Rise of Birds,* it occurred to me that to inspire such a predator to explore and learn from its complex environment, another characteristic—one of the most valued hallmarks of human intelligence, too—would prove extremely valuable: curiosity.

My hosts back at Atherton, one of them a wildlife artist and naturalist, told me they'd read a newspaper account of a researcher following a casso-wary until it sat down to rest. As the researcher waited for his study subject to rouse, he took off his backpack and leaned against a tree; soon he had dozed off in the afternoon heat. When he awoke, his pack was open and its contents strewn everywhere. In front of him was the cassowary, asleep.

Road signs near Mission Beach: entering the cassowary capital of the world

A big yellow road sign, as wide as a car, announces we are nearing our des-tination: "YOU ARE ENTERING A CASSOWARY CONSERVATION ZONE," it pro-claims. A kilometer later, an artist's rendering of a huge black bird with a blue neck and two red wattles, along with a series of eight three-toed tracks, stands above the warning: "CROSSING, NEXT 2 KM."

As we enter Mission Beach, the signs increase in frequency. They cer-tainly catch your attention. Instead of the symbol of the little girl in the old-fashioned dress flinging herself into the street, or the boy and girl with no hair and perfectly round heads walking to school, or the symbol of a truck that appears to be chased by two snakes, or the car on what looks like a slab of cheese, most of the yellow road signs here feature the image of my quarry. "CAUTION:" and beneath the word, a picture of the cassowary. A yellow diamond sign emblazoned with a cassowary, atop an orange rectangle, advises a speed of 40 kilometers per hour. Later, I would discover, along smaller roads, smaller signs that people rush out and plant when the occasion demands: "TAKE CARE"—and below it, a silhouette of a cassowary father followed by two chicks, advising "RECENT CROSSING." My favorite sign is the biggest: it's the size of a small car and shows a car that appears to be leaping into the air toward a cassowary that is just about the same size. Beneath it: "SPEEDING HAS KILLED CASSOWARIES." Another sign simply poses the question: "WHAT WOULD MISSION BEACH BE WITHOUT CASSOWARIES?"

To Mission Beach, the cassowary capital of the world, "cassowaries are really a flagship here. That's what it's all about," says David. One of the main drags is "Cassowary Drive"—which leads to Wongaling Beach Road and its small shopping center, presided over by a two-storey statue of a cassowary. Stores sell cassowary postcards, cassowary T-shirts, cassowary tote bags. Lodges and motels advertise the possibility of a cassowary sighting in their brochures. Cassowaries merit an entire paragraph in the four-paragraph introduction to the official *Mission Beach Street and Business Directory.*

Actually a conglomeration of four villages—Bingil Bay, Mission Beach proper, Wongaling Beach, and South Mission Beach—Mission Beach may not have the highest concentration of cassowaries on earth, David tells me. But it heartily celebrates the ones they have—by one recent but contested estimate, about forty adults, perhaps thirty brown "teens," and about thirty chicks.

Such public appreciation is largely thanks to an organization of area environmentalists who call themselves the Community for Coastal and Cassowary Conservation, or C4 for short. The founders of the group first formally met after a conference on the plight of the cassowary in 1991. The bird was then listed as threatened, because of habitat destruction, cars, and dogs; in 1999 it was listed as federally endangered. According to C4's attractive website, www.cassowaryconservation.asn.au, perhaps only nine hundred of the

birds live in the Wet Tropics—a figure David considers less a scientific estimate than a guess, given the difficulties of counting cassowaries. But there's
no doubt that these dinosaurs, while not extinct, are in trouble. Queensland,
the Australian state in which cassowaries are found, suffers more land clearing than any other. According to C4, Mission Beach has lost half of its cassowary habitat due to clearing.

In response to these threats to the flightless bird, cassowary fans flew into
action. The volunteers of C4 established a land gift fund to buy back critical cassowary land; lobbied leaders to reduce speed limits and post warning signs; and built a nursery to propagate seeds the volunteers collect
from cassowary droppings. They organize "plant-outs" to reforest degraded
areas, support a rehabilitation center for injured cassowaries and other
native animals, and run an attractive and informative nature center on the
area's main drag. And they lobby to protect cassowary habitat from further
development—sometimes a source of friction in a resort area.

As a scientist, David needs to distance himself from "cassowary politics,"
he tells me. But he wants me to meet one of the officers of C4, Liz Gallie.
She can give me some pointers on the best spots to see a cassowary. We stop
at her house in the woods in Bingil Bay, where once a cassowary walked into
her kitchen and stole an apple from her fruit bowl. Though she's a busy sil-
versmith rushing to leave on an out-of-town trip, Liz takes time to give me
a list of cassowary hotspots. I'll track them down once I get a rental car and
a map.

That's the next thing I do, once I check into my motel. For now David
is leaving me, rushing back to Atherton for an in-service refresher course in
first aid. I'll be alone in my quest for a dinosaur. I'll have only four more days
to find one.

I begin my day with the forty-five-minute climb up Bicton Hill, one of
the trails on my list of cassowary hotspots. The four-kilometer trail winds
through stands of box brush, swamp mahogany, milky pine, and stinging
tree, the branches draped with epiphytes thick as bunting: staghorn fern,
bird's-nest fern, elkhorn fern. The trail ascends to a spectacular lookout over
beaches, coastal plain, and banana and sugarcane plantations. Near the sum-

mit grow cycads—stout-trunked, palm-like, cone-bearing trees that evolved in the late Carboniferous, 300 million years ago; they were among the first plants to have both cones and seeds. Cycads throve so spectacularly in the Jurassic, comprising perhaps 20 percent of the flora on the face of the earth, that another name for the era is the Age of Cycads. What a good place to look for a dinosaur.

I am all eyes and ears, my senses hungry. I drink in the liquid coos of emerald dove and the languid flight of the Ming-blue Ulysses butterfly. Is that rustling sound an orange-footed scrub fowl, scratching for insects in the leaf litter? No—that is the sound of a *big* foot. I slow my pace and walk softly. I turn a corner.

I am greeted by the sight of an agile wallaby in the path. With a reddish coat and pale underparts, he stands a little over two feet tall, with a relaxed, hunched, pear-shaped posture, his clawed hands hanging in loose fists in front of him. He swivels one black-edged ear in my direction and looks at me without great surprise. When he bounces leisurely away, it feels almost like an act of courtesy, yielding the trail to me, as would a generous host.

Minutes later I alarm a five-foot-long black and yellow lace monitor, a predatory lizard as big as a young crocodile. He had been basking in the early morning sunshine. He climbs a tree to escape. Like a squirrel, he hides on the side of the trunk where he knows I can't see him. When I move, he moves; I can see only the tip of the tail here, a claw there. Monitors are the smartest of the lizards, and studies of them show they are capable of anticipating the evasive tactics of their prey. I encounter another one of these intelligent lizards just a bit farther up the trail. He is much smaller, with a kinked tail. Oddly, he seems less worried about me than about his larger neighbor. Although he, too, climbs a tree for safety, he stays nearby and lets me watch him openly, flicking his forked tongue in and out, tasting my scent.

From such an arboreal perch, many scientists believe, the early birds launched their first gliding flights. Perhaps they first climbed trees, as the monitors did, to avoid dangerous ground predators. Perhaps they found in trees an unexploited trove of insect prey. Possibly the first flight looked more like that of a flying squirrel than that of a bird. That was the case with *Microraptor*, the new species of flying dinosaur unearthed in China's Liaoning Province in 2003 that shocked the world with its *four* feathered wings, on its legs as well as its arms. Analysis of the shoulder anatomy shows it could not have flapped but almost certainly glided. Using models tested in wind tunnels, scientists have determined that *Microraptor* probably launched from a tree head forward in a dive, with legs extended straight back, forming a canopy over the tail. It would have braked by swinging its back wings beneath it, hitting the destination tree first with the back claws, then those on the forewings. But *Microraptor* was certainly not the first flier; though it may have shared a common ancestor, it is at least 20 million years younger than *Archaeopteryx*. Other paleontologists believe that flight arose instead from the ground up, as land-dwelling dinosaurs learned to run and flap their winged forelimbs until they achieved liftoff. Still others, as University of Montana biologist Kenneth Dial proposed in the journal *Science*, think the bird-dinosaurs flapped their feathered forelimbs to gain traction to run uphill—as chickens do—and used this same technique to get into trees.

Our kind may never know how the first birds flew, but the lace monitor's ancestors were there to see it. I find it hard to take my eyes off the beautiful reptile, but if I keep staring up a tree I'll never find a cassowary. When I pull my eyes back to the ground, I find a gift waiting there for me. A wet, dark, greenish pool about nine inches in diameter, it's studded with at least three different species of seeds, the largest the size of those of a pomegranate. As I resume my upward climb, I start to see cassowary scat everywhere—some of it quite recent—all full of seeds large and small. At least one cassowary was here, on this path, at this very spot, within the past twenty-four hours.

I walk, pause, listen; wait; resume. My eyes try to penetrate the rain forest's darkness, looking for movement. But I catch no glimpse of the bird who left the scats so recently. I'll return later.

Next I look for the Licuala Fan Palm Walking Track, another hotspot on my list. But failing to find it, I proceed to Lacey Creek, which like Licuala is within the Tam O'Shanter National Park, a coastal lowland vine forest that Liz had recommended. Here I wander through Alexandra palms and white bolly gums, under the shiny, dark green leaves of the beeches that struggle valiantly to reestablish the dense pre-Larry canopy. Forest kingfishers zoom like arrows as they chase dragonflies from their high perches; ornate nursery frogs issue metallic beeps. Along the short, well-maintained track I find a wonderful Cassowary Food Plant Walk, with signposts narrating each fruit's importance. Excellent displays tell about cassowaries and conservation and remind you, if you're lucky enough to spot the birds, not to feed or touch them, and, as the sign puts it, "Be Cass-o-wary." The track is promisingly studded with cassowary scat. But I see no cassowary.

Now it's past noon, and I drive to another forest walk along Lacey Creek. This one is mostly without signposts, but it claims to lead, in 4.7 kilometers, to Licuala State Forest, whose entrance I have been unable to locate by car. Much of the track here is red mud, a thin film of water coursing over it. Miraculously, in a small, irregular depression in the mud, I see tadpoles wiggling, sperm-like, in the shallows. I watch them, worried for their future. Their water world is drying before my eyes. But like their puddle, they too are changing into something else—perhaps in time to save them.

I am witnessing the story of evolution writ small. It's the story of the dinosaurs who became birds—the swift, smart predatory reptiles whose scales transformed into feathers. But why? Did feathers arise for warmth? For show, to impress rivals or the opposite sex? For protection against parasites or predators? (Like certain lizards' tails, feathers come off in the mouth of a predator, as many a fox back home knows.) Paleontologists in different camps hold fast to each theory.

Both the tadpole and the cassowary seem caught in a moment of transformation, and that moment holds me transfixed in its mystery. It's during moments like these that I time travel, back to an era of wondrous transformation, flying on the nearly vanished wings of the cassowary.

Scientists are fairly certain that the ancestors of the cassowary could fly. (Interestingly, native peoples in cassowary lands think so, too: both Aboriginal and New Guinean legends tell of flying cassowaries.) Cassowaries are members of a group of birds called paleognaths, literally "old jaw" because the palate is more primitive than that of other modern birds; paleognaths include the African ostrich, the South American rhea, the kiwi of New Zealand, and the Australian emu (the species most closely related to the cassowary), as well as the extinct thousand-pound elephant bird of Madagascar and the many extinct species of moa in New Zealand, some thirteen feet tall, all entirely flightless.

Older than the ancestors of the ducks and the buzzards, older than the birds that would become finches and flickers, chickens and crows, the pale-ognaths are thought to be the first of the still-living bird lineages to have diverged from the original bird lineage more than 100 million years ago. This was in the middle Cretaceous—long after birds had learned to fly.

Even though the oldest known cassowary fossil is merely 5.2 million years old, scientists know the bird's ancestry because it is written in genetic code. All birds in the paleognath group share very similar genes; that the giant, flightless birds should be related seems unsurprising. But the flightless giants are also genetically similar to a little-known neotropical bird called a tinamou—a grouse-like creature whose short, rounded wings enable somewhat clumsy but swift, flapping flight. New genetic evidence, published by three separate teams of authors in 2006 and 2008, reveals the tinamou is not just a distant relative that retained flight long after the appearance of the giant, earth-bound species. The genetic information shows the flying tinamou is more closely related to paleognaths of Australia and the islands of New Zealand than to Africa's ostriches and South America's rheas. The data strongly suggest that the ancestors of the paleognaths could fly and that flight was lost at least three separate times on three separate continents.

Why would a bird lose the power of flight? Native peoples have long considered this and generally conclude it to have been the result of some kind of blunder. Some Aboriginal legends suppose the early cassowary was mean and hurt children, so a hero cut off its wings. One New Guinea myth claims the first cassowary was clumsy and destroyed its wings when it crashed into a sapling. But a story among a certain group of Central Highland people in Papua New Guinea may have gotten it right: following a swiftlet's invitation, the heavy cassowary tried to perch on a twig, fell to the ground, and never flew again. In other words, the cassowary lost its wings because it was too heavy.

Scientists might well agree. Growing large could have benefitted the ancestors of the cassowary. They could swallow fruits too big for other birds to eat; their size could intimidate predators. But a bird this heavy could not hope to fly. Happily, that no longer mattered because their food falls to the ground, and they don't need to fly to flee predators. I can imagine, as in time-lapse photography, the cassowary's useless wings growing smaller and weaker over the eons, becoming stunted stumps tipped in vestigial quills, the once-complex, triple-barbed feathers simplifying into twin-shafted black bristles that hang like hair.

And now I return to the present—and suddenly see what I have been staring at these many minutes: the tadpoles' microhabitat is actually a series of three miniature finger lakes, each less than nine inches long. They are swimming in the three-toed track of a cassowary.

Frantically I search for more. In the mud, the cassowary's trail should be easy to find. But no; I find only the boot prints of other hikers and the hoofprints of feral pigs.

I push on. The day grows hotter, the trail more open. Thanks to Cyclone Larry, there is so little canopy now that I have to put on a hat to avoid a sunburn. The blood vessels in my face feel like they are exploding in the heat. My shirt is soaked with sweat, my hair wet and stuck to my face, my socks squishing in my hiking boots. My head buzzes and throbs audibly, like the voices of the singing cicadas. I realize I am getting dehydrated and take a swig of water from my canteen.

Two and a half hours later, I reach the end of the track. The trail never reentered dense forest. I saw no more tracks, no scat, and no cassowary. Rather than retrace my steps through this impoverished habitat, I hitchhike back to my parked car. The driver who picks me up, a woman about my age, recommends the Cutten Brothers Walking Track, a 1.5-kilometer trail through mangroves and lowland forest where her teenaged son had seen a cassowary on a hike two weeks ago. On the way back, the road signs seem to be taunting me: "CASSOWARY CROSSING: NEXT 2 KM." "CAUTION: CASSOWARY." "PLEASE DO NOT FEED CASSOWARIES."

But where are they?

Fortunately, over the next two days, I find that the local people are eager to help with my quest.

The kind fellow from whom I rented my car, Harold, suggests I try Nonda Street, just one street over from Cassowary Drive. "There's one in there," he assures me. "I'll see it every day for four days and then not see it again for weeks, though."

"I always see one on the Bicton Hill Circuit," the elderly lady at the convenience store tells me. "I don't go there every day, but I've always seen one when I do."

"There's a B and B that has one who visits regularly," says the helpful volunteer at the Wet Rainforest Information Center. She phones up the lodge and the owners, Mick and Sue Lukowski, generously invite me over. It's on Mission Circle, just off Cassowary Drive.

I take everyone's advice. On a rainy morning, I visit the Lukowskis at their pretty, verdant Licuala Lodge. Cassowaries feel so at home here, they tell me, that once, while Mick and Sue were carrying groceries from their car to their pantry, a cassowary stuck his head into the open trunk, sliced open a packet of Danish pastries, and ran off with the sweets, leaving the plastic behind. Now the cassowary occasionally pecks at car windows, and they have to tell guests to cover their parked cars with a sheet. But guests consider it a worthwhile trade-off. "For birders, the cassowary is their holy grail," Mick tells me as we share tea on the veranda overlooking the garden. "They come up here and if they see one, it's very exciting for them."

A male and female have been courting in the backyard for weeks. The male had followed the female, getting closer and closer, until finally, last week, he was seen preening her feathers gently with his bill. Unfortunately, the couple hasn't been seen for the last two days. "They're off in the rain forest, presumably, doing their thing," Mick says. But their absence ups the chance we'll see one of their chicks—a three-year-old the Lukowskis call Squiggles. He's got a three- or four-inch-tall casque and adult plumage, except for a bit of brown fluff remaining on the back of the head and under the tail. He's also got a limp, though no wound is apparent. Possibly, the Lukowskis think, he twisted his leg running from the parents, who chase him from the yard whenever they see him.

I spend a pleasant morning with Mick and Sue, but Squiggles eludes me. (Later that evening when I phone, I discover he came to call an hour after I left.) I drive Nonda Street regularly (it's a neighborhood, so to sit in my car would arouse suspicion), but I never see a cassowary there. I also explore the other hotspots that Liz and others have mentioned, including the Kennedy Track, a seven-kilometer trail through mangroves with magnificent views of the sea. A mangrove swamp might seem odd habitat for a cassowary, but records from the 1930s report sightings of cassowaries sitting in shallow mangrove water, letting fish swim into their long, bristle-like feathers. Then the birds, it was said, would shake the fish out and eat them. Could this be true?

A swimming cassowary bears an alarming resemblance to a plesiosaur.

I don't get to find out. I next try the short Cutten Brothers Walk, where the kind lady's son had seen a cassowary, and Ulysses Link, through rare lowland rain forest, some of the last left in Queensland at sea level. I scour the three tracks in the Licuala State Forest. And every day, starting at dawn and again at dusk, I follow a ritual: I walk slowly and quietly up to the summit of the Bicton Hill Track. I walk past the spot where I often see the wallaby, past the trees that the lace monitors climbed, past the wonderful cycads that once nourished the dinosaurs, my careful, quiet steps like a prayer: Cassowary, I'm here. Cassowary, come out. Cassowary, are you there?

On the third evening, I get an answer.

At first I think it's my drinking water sloshing in its bottle. But it continues even while I am still. It sounds like the voice of a big conga drum, but deeper: a throb so deep I cannot so much hear as feel it in my chest, not loud, but powerful. Five beats in quick succession. Then six. Then four. I can't tell where it's coming from. It seems to be coming from everywhere, like an earthquake or a distant volcano.

I have heard it before.

"When you think of *Jurassic Park,* and the sort of scenes when dinosaurs are coming towards them and they're going 'What the hell is that?'—that's sort of what it sounds like. It's like something Spielberg cooked up," Andy had told National Public Radio. And then, the show played the calls Andy had recorded, calls he later played for me in his car: resonant, stirring, thunderous.

Not everyone can hear the calls, Andy told me. But it rattles your bones and would vibrate the liquid in a coffee mug. "One time I was with a student in the field," Andy told me, "and we were close to a cassowary but couldn't see it. It boomed. The student looked at me quite startled; he said, 'Is that an earthquake?'"

It was no earthquake, but Andy's audio spectrographs show that a cassowary call has components of one: pulsed, booming notes, some of them at frequencies below the threshold of human hearing. Such sounds are known as infrasound.

Many animals, including birds, almost certainly hear infrasound; birds might even use the infrasonic rumblings of the ocean or the shifting plates of the earth to help navigate on their migrations. But with few exceptions, like prairie chickens, most birds are too small to make infrasound. It takes a great deal of effort even for a bird as large as a cassowary. In one of his scientific papers, this is how Andy describes the contortions of a captive Bennett's cassowary as it summons its amazing voice:

> Prior to booming, it made several gulping motions, possibly inflating internal air sacs. It then opened its bill wide, raised its body upward inhaling deeply, then threw its head down between its legs and began booming . . . The entire bird vibrated visibly during the booming. The colorful, wrinkled, naked skin at the back of the lower neck inflated tightly during vocalization, roughly doubling the apparent width of the neck. The heavily wrinkled skin of the neck allows for that significant expansion.

This is what a 150-pound bird is doing, possibly just yards from me, in the dark forest. But I cannot see him.

What is he saying? Is he calling for a mate? Is he intimidating rivals? Is he warning me to stay out of his territory? This I don't know. I also cannot judge how far away the cassowary is; my ears and brain were not made for this sound. But a cassowary's are, and so is its entire way of life. Such low frequencies can travel over long distances; the sound is not dampened by the thick understory of wet leaves, as are high frequencies. This is why elephants and whales also use infrasound to communicate with one another when they are far apart. One of my best friends, Liz Thomas, was a member of the team headed by Cornell researcher Katy Payne that first published on elephants' use of infrasound in 1986. They solved the mystery of why, when elephants are being killed by poachers or in "culls," other elephants, even miles away, become agitated and upset and flee. Infrasonic communication also explains how whales find one another on migrations that span the globe to sites that vary from year to year.

"Such low frequencies are probably ideal for communication among widely dispersed solitary cassowaries in dense rain forest," Andy wrote. And infrasound was probably also a perfect way for other very large individuals, living far apart from one another, in similar habitat: the dinosaurs.

Many dinosaurs—especially the duck-billed hadrosaurs—had tall, bony crests like the cassowary. One, a four-ton giant stretching thirty-five feet from nose to tail, so closely resembled a cassowary that scientists named it *Corythosaurus casuarius*. Andy showed me a specimen at the Carnegie Museum in Pittsburgh. Its skull, we read on the plaque by the fossil, bore elongated air sacs that extended into the crest, "which probably allowed it to act as a sounding device."

Sound reception is exactly what Andy thinks the cassowary's tall, mysterious crest is for. Many theories have been offered for its function—that it protects the head crashing through the forest; that it's a secondary sex characteristic; even that it's used to scrape up leaves from the forest floor when they look for fruit. But cassowaries don't use the crest as a helmet; both sexes have crests. And they certainly don't use it instead of their magnificent feet to scrape leaves: that would be like using your head, instead of a spoon, to stir your coffee. Andy dismisses these theories because he has dissected the casques of freshly dead cassowaries and found them to be much more complicated structures than hooves, horns, or antlers. Though museum specimens are hard as bone throughout, "in life," he told me, the casque is "spongy and fluid filled. In the museum, the spongy bit hardens and looks solid. And this wouldn't show up in fossils, either," he noted. "This could well be the case with the crests on the dinosaurs. Almost certainly these big dinosaurs had to be making infrasound." The helmet is a vessel for magnifying sound.

I stand so near a living dinosaur that the throb of his call, carrying across millions of years of evolution, still echoes in my chest. I hope that he will call again—and that another cassowary will answer.

I wait. I listen. But there is no more booming; no footfalls from large, scaly feet. I hear only the tingling calls of cicadas and crickets in the dark.

My last day in Australia: I wake before dawn, frantic. I leave this afternoon on a three thirty bus to Cairns and fly back to the States the next morning. Today is my last chance.

I rise at 5:00 a.m., and before dawn, I've raced two-thirds of the way up Bicton Hill, to the area where I heard the cassowary call last night. Cassowaries sleep, usually upslope, at night. They don't like to travel in the dark. I hope to be waiting for him, silent and unimposing, when he wakes. I sit on a bench along the path and listen to my heart pounding. I think I can hear myself sweat.

I summon cassowary images in my mind, trying to conjure one. I've seen the bird a few times in captivity. Not many zoos keep them, for they are so obstreperous; those that do usually offer a view of a creature angrily pacing the periphery of the fence. But I don't want those images, or the ones I've seen on the Internet, the slew of videos titled "Cassowary Attack" showing the frustrated birds throwing themselves feet-first against a fence. (There are also a surprising number of videos titled "Duck Attack.") Instead, I play in my head the scenes from stories my friends and colleagues have told me of intriguing and evocative encounters with cassowaries in the wild. I imagine Cass bending curiously over tiny, delicate Margaret, watching her weed the cassowary garden. I think of the cassowary who walked into Liz Gallie's kitchen and took an apple from her fruit bowl. I imagine the scene when David saw his first cassowary, when he was twenty-one. He was simply visiting a friend, not conducting a study. As they were drinking tea, a cassowary strolled through the yard. Later during that visit, as David was sitting by the side of a river, slapping mayflies and feeding them to fish, he saw the same cassowary again. The bird had come down through the forest—apparently, David thought, "to check me out." He followed it on and off for part of two days, deferentially walking twenty meters behind.

What magic can I summon to draw the cassowary to me? Alas, the magic belongs to the cassowary, not to me. People who live among cassowaries believe in their powers. At certain times of year, the Kalam of New Guinea won't even speak the bird's name. When they plant their taro gardens, they call the cassowary "Truly Big Man" and "Truly Important One" instead. They say the cassowary, the rain forest gardener, can protect or destroy their crops. The Tjapukai Aborigines believe that the cassowary is so primally important that creation began in one of its capacious eggs. Lightning split open the huge egg, and from it spilled the two great forces of the universe, the Wet and the Dry, from which all life would spring.

I long to return to the beginning of the world, as I had dreamed as a child: to the time when the earth was swaddled in ferns and cycads, when the dinosaurs were first learning to fly. I had hoped somehow the cassowary might take me there, if only for a moment. But no cassowary has appeared to me. I have waited all morning in vain.

One last time, and without any hope, I walk the short Rainforest Circuit at the Licuala Track. It's hot, near noon, and no reasonable bird or animal is stirring, except for the tiny skinks who streak across the track like rain and a burst of calling parakeets shimmering across the sky. My quest is over, failed. Reluctantly I drive back to the hotel to shower before the bus trip to Cairns.

I find, to my surprise, a message waiting. My bus is delayed. Previously booked passengers, an international group of young skydivers, haven't finished their dives yet. Now we'll be leaving after 5:00 p.m.

What to do with this annoying drib of time? I take my shower and pack my bags. I don't want any food and don't need any souvenirs. One last time, I walk up Bicton Hill.

I'm tired before I even reach the top. I stop at one of the benches. Dusk comes early in the rain forest, and dark is already closing in. I wait for a breeze to dry my sweat before I proceed.

And then he steps from the tangled shadows onto the path, as if through a crack in time.

Southern cassowary

Balletically, the cassowary moves forward with breathtaking grace. The tall, scaly legs seem to bend backward, and the clawed toes rise, curl, and open and fall as softly as a pianist playing legato. One hundred and fifty pounds moving over the leaf litter miraculously makes hardly a sound.

I am less than ten feet away—so close I see the two scarlet wattles swaying on his blue neck, so close I can make out three quills on his right wing, so close I see the eyelashes fringing his orange eyes. The cassowary is tall and regal, his six-inch mahogany-colored casque like a crown. The bird bends down to delicately lift an orangish fruit—the seed of a cycad?—from the ground with his beak. When he stands upright, I watch the seed slide, slowly and sumptuously, down the length of the naked, wrinkled neck. The cassowary turns his head. The casque, I now see, curves slightly to the right.

The cassowary must know I am here. His eyesight, like that of all birds, is thought to be excellent, his hearing acute. Surely he heard me coming many minutes before. Surely if he can spot a seed on the ground, he sees the 120-pound creature with yellow hair sitting ten feet away on the bench?

If so, he gives no sign. With brilliant orange eyes, the cassowary looks right through me, as if indeed I have traveled back in time: for neither I nor my species yet exists, and creatures like this one—intelligent, curious, serene—are, for now, the rightful rulers of the world.

Hummingbirds

Birds Are Made of Air

An incubator is just a glass and metal box, about the size of a double-wide microwave. Nothing much to look at: it's got a couple of dials and a glass front and makes a soft humming sound as it goes about its task of keeping the temperature inside a constant 85 degrees Fahrenheit and maintaining 45 percent humidity. But I have spent the day flying across the country, from Manchester, New Hampshire, to San Francisco, California, ridden a bus across the Golden Gate Bridge to Marin County, and finally caught a ride with my friend Brenda Sherburn, all for the chance to look in this particular machine. For today this incubator is a jewel box, containing priceless living gems.

The machine sits on a table by the window in the guest bedroom at Brenda's house. I hold my breath as she swings open the glass door. She reaches in and removes a small red plastic utility basket—the kind in which parents store their kids' crayons—and places the basket atop the incubator so I can get a better look.

There, raised an inch above the tissues lining the bottom, resting on a pedestal fashioned from the cardboard core of a toilet paper roll, is a cuplike nest the diameter of a quarter. It's soft as cotton candy, made from puffs of plant down and strands of spider silk and decorated with lichens. Inside, facing opposite directions, with short black bills and eyes tiny as dressmakers' pins, are two baby hummingbirds. Each is less than an inch and a half long. They are dazzlingly perfect, tiny, and vulnerable.

A baby in his nest

Together they weigh less than a bigger bird's single flight feather. They are probably eleven and thirteen days old. Already the greenish feathers on their heads hint at the opalescent glow that inspires naturalists to call the hummingbird "a glittering fragment of the rainbow," "a breathing gem," "a magic carbuncle of flaming fire." A week ago, these birds were the size of bumblebees, pink, blind, and naked. After their mother disappeared, they spent a night and a morning alone, starving. It's a marvel they're alive.

"When I got them, I was pretty sure at least one wouldn't make it," Brenda whispers. They were then about three and five days old, with no feathers or eyesight—just hunger and need. "The first three days with them, when they're this small, it's really touch and go," she said. "When you get a baby, you don't really know what it went through till it got to you."

But Brenda knows exactly what the babies need right now: two hundred fruit flies. They're best caught fresh, crushed with a mortar and pestle, then mixed with a special nectar supplemented with vitamins, enzymes, and oils. From dawn to dusk, this food must be delivered into the babies' desperately gaping mouths by syringe, every twenty minutes. Because the food spoils easily, a fresh batch must be concocted several times a day. Brenda is one of a small handful of volunteer wildlife rehabilitators willing and qualified to do so.

I was lucky enough to meet her through a mutual friend, while Brenda was visiting family in New Hampshire the autumn before. Instantly, I liked her. A fifty-six-year-old sculptor and mother of three, she's a five-foot-three powerhouse in dark bangs and a pageboy: her skills range from casting her own bronze to founding an art collaborative for kids to raise funds for conservation. But perhaps even rarer than her energy is her patience. As we sipped tea in my kitchen the afternoon we met, she answered my questions for more than two hours, speaking in careful, sometimes halting, always thoughtful phrases of the intricacies of hummingbird rehabilitation. The work demands extraordinary precision and commitment. The plights of her charges are often pathetic. Yet, as she spoke, her brown eyes shone with merriment. "The word 'cute' was really invented for a baby hummingbird!" she said. "They are so cute and so fast, so curious and smart—and yet so little is known about them.

"To put a little hummingbird back in the wild," she told me, "might seem like a little thing. But it's a big thing." What is it like to restore these tiny glimmers of birds to the sky? I wondered. Brenda was generous enough to invite me to come to California to the home she shares with her partner, Russ La Belle, to observe her work.

Most veterinarians will tell you that orphaned and injured birds are far more difficult to treat than mammals. In birds, often the first symptom of illness is death. Birds are physiologically very different from us. Humans and our fellow mammals are fluid-filled creatures. Early Greek physicians believed all medicine could be based on an understanding of these fluids, which they called humors. But birds, in order to be freed for flight, cannot afford to be loaded down with heavy fluids. Birds are made of air.

Unlike our thick, marrow-filled bones, most birds' bones are hollow. Even their skulls are scaffolded with passageways for air. Their feather shafts are hollow, and the feathers themselves, like strips of Velcro, are interlocking barbules for catching air. Their bodies are filled with air sacs, which originate in, and function, in part, as extensions of the lungs. No fewer than nine of these filmy bladders fill the tiny body of a hummingbird: one pair in the chest cavity; another under each shoulder blade; another pair in the abdomen; one under each wing; and one along the neck.

Hummingbirds are the lightest birds in the sky. Of their roughly 240 species, all confined to the Western Hemisphere, the largest, an Andean "giant," is only eight inches long; the smallest, the bee hummingbird of Cuba, is just over two inches long and weighs a single gram.

Delicacy is the trade-off that hummingbirds have made for their unrivaled powers of flight. Alone among birds, they can hover, fly backward, even fly upside down. For such small birds, their speed is astonishing: in his courtship display to impress a female, a male Allen's hummingbird, for instance, can dive out of the sky reaching sixty-one miles per hour, plunging from fifty feet at a rate of more than sixty feet per second—and pulling out of his plunge, he experiences more than nine times the force of gravity. (Adjusted for body length, the Allen's is the fastest bird in the world. Diving at 385 body lengths per second, this hummer beats the peregrine falcon's dives at 200 body lengths per second—and even bests the space shuttle as it screams down through the atmosphere at 207 body lengths per second.)

Hummingbirds' wings beat at a rate that makes them a blur to human eyes, more than sixty times a second. For centuries, people deemed hummingbird flight pure magic. Until the invention of the stroboscope, scientists could not understand how hummingbirds hover. With a flash duration of one hundred-thousandth of a second, the stroboscope finally revealed the motion of wings that had been too fast for other cameras to capture.

Hummingbirds are less flesh than fairies. They are little more than bubbles fringed with iridescent feathers—air wrapped in light. No wonder even experts who are experienced with other birds are intimidated by this fragility. "Their feet are like thread," another rehabilitator, Mary Birney, who lives in Pennsylvania, told me. "Touching them damages their feathers. Yes, they are made of air—air and a humongous heart. That's all they are. It floors me I'm able to work with them."

Hummingbird rehabilitators are unsung heroes. Toiling away with their syringes and tissues, each is a Mother Teresa, a Saint George, a little Dutch boy with his finger in the dyke—desperately trying to fend off the hoards of monstrous perils facing these tiniest of all birds. Hawks, roadrunners, crows, jays, squirrels, opossums, raccoons—even dragonflies and praying mantids—eat them. Bass leap from ponds to gulp them whole. Fire ants and yellow jackets sting babies to death in the nest. Flying adults get impaled on the stamens of thistles. They are killed by unseasonable freezes—and by other hummingbirds. They spar with needle-like bills, but most hummers kill rivals by chasing them away from nectar sources. The losers starve.

They die from infestations of mites. They get blown off course on migration and run out of energy. They fly into spiderwebs while hunting for bugs, or while gathering the silk for nest making. They fall to the ground with their wings bound, mummy-like, in sheets of sticky silk, unable to fly or feed. One woman found such a victim on the floor of her barn, so dirty and lifeless-looking that she kicked it with her shoe before realizing it was not a clod of dirt but a glittering, still-living hummingbird, imprisoned in a robe of cobwebs.

The hummingbird's world becomes yet more hazardous with human disturbance, especially the relentless destruction of swamps and woodlands—hummingbirds' best nesting areas, full of nourishing bugs as well as energy-giving flowers. We kill these beautiful birds, too, with our pets. The most common reason for any bird's admittance to a wildlife rehab center is abbreviated on forms as "CBC": Caught by Cat. Hummingbirds also smack into our windows and are hit by our cars. (They are so attracted to red they may kill themselves hitting the glass pane separating them from a red flower in a greenhouse. They will chase red cars. A woman with a scarlet Chevy Suburban reported that immature hummingbirds would probe for nectar in the cracks of its hood.) And because they are all lung and so small, hummingbirds are extremely vulnerable to our pollutants and poisons. Brenda has seen all too many hummingbirds poisoned by common garden pesticides, for which there is no antidote.

Once Mary Birney, a former pharmacy technician with University of Pennsylvania, treated an adult ruby-throated hummingbird who had been trapped in a garage and gotten stuck in spilled polyurethane. The homeowner brought the miserable victim to an animal rescue center. Volunteers used canola oil to remove the sticky goo, but then the bird was oiled, like some poor cormorant after Exxon *Valdez*! A staff veterinarian tried to wash off the oil, but the force of the water coming from the faucet ruptured one of the bird's air sacs. By the time Mary picked up the hummingbird to bring it home, it had swelled up like a toad with subcutaneous emphysema. Mary used a 60 cc syringe to gently wash the bird again—and even this ruptured another air sac. Yet, in the face of all these horrors, Mary observed, "hummingbirds have such an amazing life force." After keeping it for weeks in her home, she released it, healed, to the wild.

Alas, successes with injured hummingbirds are rare. Most medical treatments kill them. Broken wings or legs can't be splinted—a splint would be too heavy (though Brenda, with help from a mentor, once successfully splinted a hummer's toes with a strip of stiff paper). Injections risk puncturing an air sac. And to a bird who weighs as little as a penny, even the slightest overdose of medicine can be lethal. The survival rate for birds admitted to WildCare, the wildlife rehab center with which Brenda works, is impressive: the last year for which stats were available showed that 46 percent were saved (only slightly lower than the 54 percent survival rate for mammals). But that's not representative of sick or injured adult hummers. Almost all of them die.

Sometimes kindhearted people do orphaned and injured hummingbirds more harm than good. One would-be rescuer had fed a baby hummingbird lentil soup and managed to coat its feathers in the bargain. Another gave a hummingbird yogurt and honey (both of which could be lethal). One woman brought in a bird in convulsions. The hummingbird had hit a window; she had tried to treat it with a drug from her own medicine cabinet. Now it was suffering from an overdose of Valium.

Too often, people "rescue" babies prematurely. "People find a hummingbird nest and panic," says Brenda. If the babies are alone, they assume the nest is abandoned. But to feed her young, a mother hummingbird must leave the nest 10 to 110 times a day. WildCare staff always tell callers to watch the nest for at least twenty minutes to make sure the babies are really abandoned. "The problem is," Brenda says, "so few people will sit still and just watch something for that long." Invariably their eyes wander from the nest and they miss the mother's lightning-quick return. Some callers won't take the time to watch at all. They'd rather pluck the nest from the tree and take it to WildCare than be twenty minutes late to work.

Orphaned babies do much better than injured adults, though. They have an excellent chance with Brenda, who has been working with hummingbirds since 2001 and apprenticed for a year with her mentor, a professional veterinary assistant. And unlike in the East, with its single hummingbird species, the ruby-throated, the San Francisco Bay Area hosts four hummingbird species. Brenda has worked with them all. Black-chins, with their shiny dark heads, white collars, and violet neck bands, migrate through the area. The rufous, with its emerald crown and reddish back spangled with green sequins, passes through on its astonishing trip from Mexico to nest in Alaska—the longest migration of any bird on earth in terms of body lengths. Anna's hummingbirds, named in honor of Anna Massena, duchess of Rivoli, live year-round in Brenda's garden; the males' throats glow like "roses steeped in liquid fire," as one writer put it. And Allen's hummingbirds, looking very much like the rufous, also nest in Brenda's neighborhood. In fact, the species was named after bird collector Charles Allen, who also lived in the Bay Area. Some hummingbirds here may start nesting as early as February; some may raise second clutches as late as August.

So when Brenda phoned me on a Wednesday in June to say that she had received a pair of orphaned twins, I booked my flight immediately. She was offering me an extraordinary opportunity: the chance to watch these most gossamer of birds grow into masters of the air.

At first, in their exquisite nest, the babies seem as still as stones in a jeweled ring. But suddenly, with shocking speed—if I had blinked I would have missed it—the larger one whirls in the nest to face the opposite direction. Both babies now face south. In a trice, the other baby spins in response. Both snuggle featherless breasts into the nest's soft down lining. Because it's woven with spider silk, the nest stretches to fit new positions precisely. The babies go still again, but their black eyes are open and full of life.

When they came to WildCare, they were nearly dead. "They were not moving or responding," Brenda said. "They were nudies"—without feathers—"and so dehydrated, their eyes, even though they were still closed, seemed to sink into their heads. They looked like little skeletons." WildCare staff put them in a closed oxygen cage; Brenda came to pick them up within an hour. She feared they might not survive the fifteen-minute drive back to her house—particularly the smaller, weaker one.

Most species of hummingbirds usually lay two eggs, which hatch two days apart. The developmental difference between twins can be striking. The larger one is sleeker, fatter, faster, stronger. Its bill is perhaps an eighth of an inch longer. The younger one looks a bit rumpled, with a puff of down still on its head. Both still have a smattering of pinfeathers. Sheathed in a protective wax, the pinfeathers will eventually blossom like flowers, but for now they look like the quills of a miniature porcupine.

The birds grow fast. "Everything's accelerated with hummingbirds," Brenda explains. If the mother has chosen a particularly good nest site, near rich sources of bugs and nectar, her brood can transform from eggs the size of navy beans to flying hummingbirds in as little as two weeks. Offspring from sites farther from food sources might take three times as long to fledge. "It takes us longer, because we are not mother hummingbirds," Brenda explains. "If we do our job well, we can get babies fledged within a month. Every day, you see them grow." In the week she has known them, they have transformed before her eyes. When they arrived, a novice could not have guessed they were hummingbirds: they were pink blobs, with yellow beaks so stubby they were barely visible. But now they are looking more and more like the astonishing birds they will, with luck, become.

Only yesterday they revealed their species: the rufous color appeared on their backs, identifying them as Allen's hummingbirds. The similarly colored rufous hummers had already flown north. A migratory species that summers only along the coast of California, the Allen's is known for its curiosity and interest in humans. "We're so lucky they're Allen's!" she said. "They're endangered—so every one of them we save—"

As if some inner alarm has rung, Brenda glances at her watch, then checks the timer she has clipped to her waistband. Often, she needs neither: she just knows. "Oh!" she says. "It's almost time."

She bends toward the babies, puffs her cheeks, and gently blows on them. They know this signal well: before their eyes had opened, before the mysterious disaster took their mother away, a soft whoosh of wings as the female landed on the nest promised a meal coming. In response, each gaped its stubby bill to receive her needlelike beak down its tender throat like a sword swallower.

This they do now, with furious enthusiasm. Brenda grabs the 1 cc syringe that has been sitting in a dish next to the incubator. It's fitted with a slender catheter instead of a 24 gauge needle, loaded with fruit-fly-nectar cocktail. The little beaks pop open, and they wave their heads frantically, like an over-eager schoolchild waves his arm in the air to attract the attention of the teacher. Brenda thrusts the catheter into first one throat, then the other, pushing the plunger to dispense 0.01 cc into each.

The feeding inspires the older bird to a spectacular feat. He wiggles back-ward in the nest, jouncing the smaller sibling, nearly executing a headstand. He pushes his tail high above the nest's rim—and, with impressive force, blasts out two turds, each about the size a mouse might leave. They fly more than thirty times the length of his body, past the rim of the nest, past the basket, past the table, and splatter on the window. To my horror, they look like clots of blood. Does the baby have bloody diarrhea? Brenda, calm as still water, is unperturbed.

Her husband, Russ, is blessed with a brother who "has a great compost pile," she explains serenely. "Russ's big job is netting the fruit flies fresh." He does so two or three times a day. One can order frozen fruit flies from suppliers—their usual market is zoos and reptile hobbyists. "But these are bloodier and yummier," Brenda says. "You can tell they like these better. They practically lick their chops! They're such little dinosaurs."

Indeed, the babies do seem to have enjoyed their meal. Why not feed them more?

"Because you can pop them," Brenda says. Miss a feeding by twenty min-utes and they can expire. Feed them too much and they can explode.

Brenda resets the timer.

"When I have the babies, we always have to eat something quick for din-ner," Brenda explains. Twice while we are preparing the meal (sushi, salad, bread, tortellini with pesto) and twice while we are eating it, the timer goes off. Immediately she goes to the birds to feed them. "Everything," she says, "stops for the hummers."

This sort of schedule makes it difficult to find reliable volunteers to foster baby hummingbirds. "I've had people who said they wanted to—and it's a lot of training," she tells me. "But once they finally get a baby, they realize this is *every day*. After a couple of days, they don't want to get up at five in the morning to get things ready; they don't want to change their schedules at night; they don't want to feed the bird every twenty minutes. They start slacking off. And you can't do this with a baby hummingbird."

Once, she had an apprentice who was raising an orphan. Brenda dropped by her house one day to bring her some supplies. The woman wasn't home. "So I waited for her," Brenda remembers. "I assumed she had taken the bird with her. She came driving back and was surprised to see me. I said to her, 'I wanted to give you this stuff for the bird—and where is the bird?' It's in the house—and the poor little thing is almost dead." She took the baby, and under Brenda's care, it survived.

As I watch the twins, I wonder if I could commit to such a schedule, dividing each day into twenty-minute segments, crushing bugs and mixing fresh nectar, day after day. The younger one stretches out a wing and attempts to preen, revealing the membranous skin. Tissue paper seems like armor in comparison. The beaks, too, are extremely fragile, soft and growing. In *The Secret Garden*, Frances Hodgson Burnett wrote of the "immense, tender, terrible, heartbreaking beauty" of eggs. But an eggshell is a fortress compared to the feathers and gossamer skin that shield this minute nestling from all the evils of the world. And yet—now it stands in the nest and tests its wings, whirring with a concentrated ferocity. The wings beat so fast that when we take a picture with Brenda's digital camera, the photo shows a bird with mist for wings. How can someone so fragile be so fearless? How can it summon a resting heart rate of five hundred beats a minute, revving to fifteen hundred a minute when, one day, it flies?

These little bubbles of spunk inspire extraordinary tenderness. One autumn, a ruby-throat, on its lonely, five-hundred-mile migration—a journey across the Gulf of Mexico, which can demand twenty-one hours of nonstop flight—landed, spent, on a drilling platform on the Mississippi coast. It was too exhausted to continue. The oil company dispatched a helicopter to fly it to shore. The hummingbird spent the winter in a gardener's greenhouse, then left, fat and healthy, on its spring migration. Hummingbird rehabbers told me other tales of devotion from ordinary people: folks who find an injured or orphaned hummer are often willing to drive for hours to make it to the rehab center. And now I know for sure: I would do anything—anything—to protect these precious infants who have fallen into our care.

Tonight, as the baby Allen's sleep in the incubator, my nest is a futon just yards away. Their bedroom is my bedroom, and their schedule will become mine: once darkness falls, I won't turn on a light, to avoid disturbing their sleep. I'll carry a flashlight to navigate down the hall to use the bathroom. If I want to read, Brenda offers kindly, I can do so in the living room, but tonight I don't want to. I feel full of my last sight of them, as Brenda replaced their red basket in the incubator after their last feeding, at 8:32 p.m. In their soft nest, they sleep in the shape of a smile: tails curled, backs arched, beaks facing upward toward a sky they will one day own. I almost can't wait for dawn.

5:30 a.m.: In her pink bathrobe, Brenda wakes me with quiet taps on the door. I've woken many times during the night. Fighting the urge to check on the babies, I've been waiting for this moment, when she takes the babies from the incubator to place their basket by the window, so they can wake to natural light. But when Brenda opens the incubator, there is a shocking surprise: the older twin has pushed its smaller sibling out of the nest! I feel sick at the thought of the little bird tossed from its thistledown nest, alone and frightened on the bottom of the basket for who knows how many hours during the night. The baby sits stunned but safe on the tissue-lined floor of the red basket.

"Probably this happens a lot of the time in the wild," Brenda says. Unfortunately, when people find baby birds in this predicament, they often hesitate to replace them in the nest; they're afraid to touch them, for fear a human smell will linger and repel the parents. This is patently untrue. Except for New World vultures, most birds have little sense of smell, if any. Most parents greet a baby returned to their nest with visible joy.

Brenda replaces the younger twin in the nest; to my astonished delight, she places the other in my cupped hand. "I'm putting you in a separate nest because you are naughty!" Brenda says jokingly to the older twin. She has a supply of extra hummingbird nests she has saved from previous orphans and goes off to the living room to choose one. Holding the baby, I try to sense the weight and shape of his belly, but I can't. He's as light as a sunbeam, and the skin on my palm feels thick as a shingle. I am a lumbering monster cradling a single breath in my clumsy, oversized hands.

Brenda returns with an Anna's nest for the older twin. It's lined with feathers as well as plant down, and while it, too, is decorated with lichens for camouflage, it has a greener cast to the outside. Brenda sets a small perch, a twig from one of Russ's fruit trees, on either side of the new nest, should the older bird graduate to this new skill. She sets the younger baby in the old nest just inches from the new one so the siblings can see each other and gently places the older baby inside.

The bigger twin immediately spins, shaping the new nest to his body. The younger one, now alone in the original, stretches out, rather like a person in a Jacuzzi: ahhhh! "I've never seen a hummingbird actually lounging in a nest before!" says Brenda. With both babies settled, Brenda and I move on to Phase I of breakfast: grinding coffee and crushing fruit flies.

Our day is ruled by inch-and-a-half-long birds. During the daylight hours, there is no possibility of working on a sculpture, or shopping at the grocery, or visiting a friend's house, or a workout at the gym. Brenda can't paint, or glaze her clay or ceramic sculptures, or work with pastels. For the weeks before her babies are weaned, she saves stacks of wildlife magazines to read. She sketches ideas for new sculptures. In twenty-minute snippets, she works on Saving the World. (The children's art cooperative she founded, Saving the World One Drawing at a Time, elicits drawings of wildlife from children, which are sold on her website, saveworlddraw.org, to benefit conservation projects the children choose.) Every twenty minutes, a baby hummingbird is hungry. What could be more important than that?

Russ, too, submits to the tyranny of the nestlings. He dashes to the store for the supersized Kleenex that line the hummingbirds' basket, trolls compost piles for fruit flies, picks up take-out for dinner. Everywhere his lanky six-foot figure is followed by four rescued dogs. Foxy, a white shepherd mix, has lived with the couple for most of her twelve years, but the other three dogs—Scout, a ninety-pound black shepherd mix, Star, a black lab mix with a white spot on her chest and a limp, and Buddy, a rambunctious white lab mix—are recent acquisitions. They came to live with Brenda when Russ was stationed at Ft. Leavenworth, Kansas, last year. Just months from his sixtieth birthday, the Army Reserve had called him to serve. Within the first two months of his year-long residency, Russ sent home six homeless dogs he rescued from death row. He found homes for the three he and Brenda did not themselves adopt and founded a campaign to "Save the Dogs of Ft. Leavenworth." Seven years ago, he rescued homeless bees who were living in a cardboard box at the edge of a mall parking lot in San Rafael. When Russ saw them, he went to his parents' house, borrowed his father's bee suit, and brought the bees home. Now they live in a chest of drawers in the backyard.

Russ is retired; Brenda is learning GPS mapping to supplement the couple's income. But for both of them, it's clear that their real work in the world involves more than making a salary. Fostering baby hummingbirds grew out of a volunteer stint Brenda had completed in Belize. After a hurricane had destroyed an Audubon environmental education center, she designed and built a new one. Russ raised funds from the States when the money ran out. When Brenda came home, she visited WildCare, where she had volunteered before, and ran into a rehabilitator who asked her to help with the baby hummingbirds. Facing her fiftieth birthday, which was just after 9/11 had turned the mood of so many Americans bitter and helpless, she began saving the lives of the tiny creatures that Spaniards called resurrection birds. The early Spanish visitors to the New World believed that hummingbirds died nightly and revived again. Surely, they must have thought, something that glittered so brightly was made new each day.

This is the gift these baby hummingbirds offer us: a hand in resurrection. Our lives do not stop for them; they begin again. Every twenty minutes, the birds' appetites call us, from our busy to-do lists and fast-paced schedules, back to life.

The nestlings are so minute that Brenda keeps a magnifying glass beside the incubator. Peering through this looking glass, we enter a world in miniature. Each of our tiny birds is a landscape for creatures even smaller than they.

At first it seems a speck is stuck on the beak of the smaller bird. But with the magnifying glass, I can see the speck moving. And now I can see another.

Hummingbird babies are so small that Brenda must
examine them with a magnifying glass.

Brenda knows what these are and assures me they're harmless. These are
creatures whose only habitats are the petals of flowers—and the beaks and
nostrils of hummingbirds. Unlike other mites, which suck blood, these sub-
sist on flower pollen. But when a flower's pollen runs out, the creature needs
a lift to a new blossom. The mite hitches a ride on the beak of a hummingbird, and at the next visit to a flower rushes out to its new home.

"Look—I'll show you," Brenda offers. We grab a red penstemon from
a vase in the kitchen—among hummingbirds' favorite blooms. Gently
Brenda pushes it toward the face of the little hummer, so that his beak fits
inside the flower's tube. I watch through the magnifying glass. When she
pulls away the flower, voilà! The mite has left the beak and moved to the
petals. This transfer, I later read, is made at a speed comparable to that of a
racing cheetah.

Beak mites seldom venture onto the feathers, but they remind Brenda
to check both birds for other possible mites. With a moistened Q-tip, she
strokes the feathers gently in the direction in which they are growing, then

ruffles them in the opposite direction to view the skin beneath. No mites. But the view through the magnifying glass is revelatory. Each tiny feather— the largest of which is scarcely a quarter inch long—is almost a world in itself.

Feathers are among the most complex structural organs found in nature. Nothing of comparable dimension is stronger. They are made of keratin, the same as a human's fingernails, a horse's hooves, and a rhino's horn—but the keratin in feathers, due to a difference in molecular structure, is even tougher.

A typical bird's feathers outweigh its skeleton. Feathers define a bird. By trapping and moving air, feathers protect the bird from cold and wet, and they enable it to fly. But each feather is, itself, largely air, with a stiff central shaft that is light and hollow and attaches, beneath the skin, to a muscle. Like each scale on a reptile, each feather on a bird can be raised or lowered as needed. The shaft divides the feather into two broad vanes on each side, which consist of parallel branches called barbs. At right angles to the barbs are interlocking branchlets called barbules. Hundreds of pairs of tiny barbules on each barb fit together like tiny strips of Velcro and give the feather its web-like quality. The barbules have small hook-like processes called barbicels. And on the barbicels of some feathers are even tinier branchlets, microscopic hamuli, which allow even more air to be trapped in the feather. When birds preen—running their beaks through their feathers, as these babies do after Brenda ruffles them—they are rezipping the Velcro of these minute connections.

Caught in the feathers, air gives birds their warmth and their flight; in hummingbirds, air even gives them their color. Their jewel-like radiance— emerald, ruby, amethyst—comes not from pigment, as in most birds' feathers, but from air. Except for the flight feathers on wings and tail, the top third of hummingbird feathers lack barbicels and hamuli. Instead, they have elliptical structures called platelets (utterly different from the clotting cells in our blood of the same name) filled with microscopic air bubbles. The shape and thickness of the platelet and the amount of air determine the color seen. These air bubbles diffract light into colors that reflect back in a flash of iridescence.

The color on the throat, or gorget, and head of a male is particularly spectacular. The platelets on these feathers are like flat mirrors, and light reflects in only one direction. This is why the gorget of a male ruby-throat, for instance, dazzles in sunlight but may look black in shade. Hummingbirds know this. By carefully adjusting his position in relation to the observer and the sun, a male purposely flashes his colors to intimidate a rival or attract a mate.

Today our babies only hint at the colors they may one day command. Young hummingbirds look like females, whose drabber plumage helps hide the nest in the shadows. If one or both of these babies are male—and this we won't know before they leave us for Mexico—the gorget color won't develop till spring. But these babies may grow one day to master the very sun.

Such sophisticated strategies seem far in the future. For now, it is thrilling enough to witness ordinary miracles. Like the one we see this morning at ten fifteen.

After the feeding, Brenda removes the catheter from the syringe. Instead of squirting the food down the bird's throat, she pushes the needleless, food-filled syringe onto the bird's beak. The beak, thin as a straw, easily fits into the opening. The bird's throat flutters and his black eyes seem to widen. What has happened? Brenda answers: "He's just discovered his tongue!"

The appendage is translucent, thin as embroidery thread, and extends and retracts so quickly that if we were not watching the nest with a magnifying glass we might never see it. At a feeder, a hummingbird extrudes and withdraws the tongue thirteen times a second. Many people wrongly think a hummingbird's tongue is hollow like a straw or sticky like flypaper. But hummingbirds do not sip nectar; they lap it. The tongue is forked, like a snake's, with absorbent fringes along the edge of each fork, and grooved down the center to withdraw extra nectar through capillary action. The tongue is so long that, when retracted, it extends back to the rear of the skull and then curls around to lie on top of the skull. (Some woodpeckers, too, have very long tongues—sometimes more than three times the length of the bill—demanding an equally odd storage arrangement when the organ is not in use prying insects from deep holes. In the case of the hairy woodpecker, the tongue forks in the throat, goes below the base of the jaws, wraps behind and then over the top of the woodpecker's skull, and comes to rest inside the bony orbit behind the eyeball.) With this extraordinary appendage, a hummer can drink its own weight in nectar in a single visit to a feeder.

As human babies explore with their mouths and later their hands, hummer babies explore with their tongues. Curiosity rewards a migratory species whose life will depend on finding nectar from hundreds of different plants. The smaller bird picks this moment to rev his wings—first just two seconds, then three, then five—as if giving his sibling a standing ovation for his accomplishment.

From now on, we see the questing tongue often. Through the lengthening bill, the bird extrudes the tongue three times, four times in succession. The baby's feathered face can be remarkably expressive: it is wearing what can only be described as a quizzical look: "What's this? And this? And *this*?"

This slender, silent tongue speaks of hunger for the world outside the incubator. Brenda takes the babies outside. "We're going for a walk," she tells them. Figuratively, of course: we do the walking, carrying the babies in their separate nests in their red basket. The moment we step into the sunshine, they rise from their nests and rev their wings in unison.

We thread along the path through Brenda's four-tiered hummingbird garden. She has planted all the hummers' favorites here: tubular red and yellow columbine, tall hollyhock, crimson fuchsias and salvias, orange lion's mane, dainty coral bells, penstemon, sticky monkey, gooseberry, and currant. The garden is peppered, too, with hanging hummingbird feeders and shallow birdbaths and Brenda's sculptures. Two statues were made from road-killed deer. In the garden, they are resurrected, their skeletons partially clothed in tufa, found stone, and concrete. We bathe the babies in the garden's light and shadow.

We bathe them in sound as well. From the seed feeders on the deck come the euphonious calls of chickadees, the bell-like trill of the dark-eyed juncos, the down-slurred whistle of the titmice, the "ank-ank" of the nuthatches, the "zree" of the house finches, and the coo of doves; from the nectar feeders and flowers, the whirr of hummingbird wings. The babies cock their heads and listen. The entire world is new, and their awakening senses are hungry for all of it.

I feel like a new mother with her child in a pram. *Here is the sweet, green world!* my heart silently promises the babies. *And one day, all of it will be yours.*

* * *

By 6:50 a.m. on Wednesday, our third day together, we're already on the third feeding of the morning. "The little one is really looking like a little Allen's," Brenda observes as she looks at him through the magnifying glass. Almost all the pinfeathers have blossomed into green iridescence on the back, with hints of orange on the head and tail. The bigger one extrudes his long tongue toward this strange big eye, the magnifying glass, wondering what it is.

The glass shows a worrisome development. "I see a lot more mites today," says Brenda. "Especially on the little one."

At 7:14 a.m. we do another beak-to-flower transfer. More than a dozen mites rush off the small one's beak to the flower, but others seem quite content to remain on the baby's head. Both birds are not only preening; now they're scratching themselves with their feet—a lot. Through the glass, we see that there are at least two different kinds of mites on the birds: some are tan, others reddish. Some may be beak mites—but others are clearly not.

More than twenty-five thousand species of mites afflict the world's birds. Some live in the air sacs of canaries, looking like specks of pepper. Others burrow inside the faces and legs of parrots. Chiggers, which also plague humans, beset turkeys and chickens. With piercing mouthparts like those of their relatives, the ticks, the mites feed on the host bird's blood. They itch, wreck sleep, and can cause feather damage as the bird desperately scratches to relieve the itching. Severely infested birds can develop sores, lose weight, and die from anemia.

"That may be why the bigger one kicked the littler one out of the nest," Brenda says. Since mothers reuse the old nest if they raise two broods, nest mites are common in a season's second clutch. Brenda swabs both babies with a moist Q-tip. She picks up five or six mites with each stroke and scrapes them off onto a tissue, which we take to the garbage can in the garage. "These mites can multiply really fast," Brenda says ominously. "I've seen babies totally covered. It can kill them." She falls silent. I can now count more than thirty on the little one, coursing like corpuscles over and under the feathers. He scratches his head with a thread-thin foot and shakes it. They are probably crawling in his ears. It would drive me mad.

"What can we do about this?" I ask Brenda. "Isn't there some insecticide . . . ?"

To Brenda, the idea is almost unthinkable. "Oh! I really don't want to treat him for mites. The stuff could kill him. He's so little . . ."

We try interim measures: While I hold the smaller bird in my hand, Brenda microwaves the original nest. We change the tissue lining the basket. We keep swabbing the babies to remove mites.

But it's a losing battle.

"This is a terrible choice," Brenda says. "The mites can kill them. Or we can kill them, trying to get rid of the mites. They could die from the poison. They could die from the stress. It's so scary!" But it's clear what we have to do.

We put it off as long as we can. We wait till Brenda's new apprentice, Julie Hanft, arrives. A ponytailed mother of a preschool son, she's an eager and competent volunteer who has already worked with WildCare to educate people about coyotes. We wait till the day warms up, so there is less chance the babies will get a chill. Finally, at 10:59, Brenda lets me give the babies one last feeding, while she sets up the kitchen for the treatment. I reset the timer, but they probably won't want to eat afterward.

Brenda covers the kitchen table with paper towels, like a nurse draping the surfaces of an operating room with sterile cloths. She sets out plastic surgical gloves; a clean, empty dog bowl; another clean dog bowl filled with warm water; a pile of Q-tips; a tiny stoppered glass vial of insecticide.

"They won't like it," Brenda announces. "They don't like being out of the nest. Everything is going to be traumatic for them. We'll try to do everything quickly." We steel ourselves for what's ahead.

This is why wildlife rehabilitators seldom name their charges. They'll tell you it's unprofessional. Of course Brenda knows that these babies aren't hers to name, that they aren't pets. But the real reason we haven't named these birds is this: the looming possibility they will die and break our hearts.

Brenda removes the larger one from the nest. Though they have much to learn, nestlings already know one thing: at all costs, stay in the nest. The baby struggles, ferociously clutching the nest's down lining with feet so delicate I fear they'll be pulled from his body. Finally Brenda gets him free. Julie microwaves the nest. When we remove it, a dozen mites lie dead on the paper towel.

Brenda dons surgical gloves, rolls a Q-tip in the mite powder, then strikes the swab on the lip of the vial. Excess powder puffs off like smoke. Brenda places the baby in the empty dog dish on a tissue. She narrates to Julie while she works: "Roll the Q-tip over the bird's feathers, around the neck, up the belly . . . You have to be so careful! This can kill them! . . . Now to the top of the head, now under the beak, and on each side." When she is done, the bird sits in the green dog bowl for a second, innocently stunned by this senseless horror perpetrated by those he trusts. And now for the worst part.

Removing the gloves, Brenda cups the baby in her hand, then dunks him into the bowl of water. He struggles and peeps, as if begging for mercy. "Oh, I'm sorry, I'm so sorry!" Brenda whispers. She takes him from the water, dunks the Q-tip, and wipes the feathers. She dunks the bird again, up to his eyes. The baby looks like his whole body is crying. Wet and bedraggled, the chick seems to have shrunk by a third—a wraith wrapped in wet feathers, a withered leaf plastered to a rain-slicked sidewalk. Brenda hands him to Julie, who ferries him in his nest to the incubator.

The smaller baby is even more pathetic. He struggles weakly when Brenda removes him from the nest. He, too, cries in the water. His tiny chest heaves with fear as Brenda carries him back to the incubator. This is traumatic and exhausting for everybody.

At 11:14, the timer rings. We finished the whole operation in twenty minutes. But neither bird is hungry. "We'll just let them sit quietly and hope they recover," says Brenda. We leave them in peace, setting the timer to check on them and see if we can induce them to eat.

11:47 a.m.: The larger one perches on the edge of the new nest, laboring to breathe, scratching periodically. The smaller one convulsively shakes his head and wings. Both refuse to eat.

12:13 p.m.: The larger bird seems to be settling down. The smaller bird hunches, eyes closed. We can still see the bugs crawling on him. He shakes his head violently, almost convulsively. Brenda reluctantly decides to powder and dip him again. "I am so, so sorry!" she tells him, voice cracking with emotion. "I had one die after doing this," she tells us. "He was really infested, with a dark ring that fell off by the hundreds. He was an Anna's. Oh, I hate this."

12:21 p.m.: After the second immersion, Brenda again offers both birds food; the larger bird accepts a tiny squirt and then turns away, as if the food tasted bad or made him feel sick. The other refuses.

12:45 p.m.: "Want some? Want some?" Brenda blows on the babies and waves the syringe. The larger one gapes enthusiastically and swallows his first real meal in an hour and forty-five minutes. "Yeah, you look good!" she says to him. "He's just a little older, and his tolerance . . . plus he didn't have as many bugs." He rises and beats his wings, looking strong as a crowing cock. The food has revived him. This bird is going to make it.

But the smaller bird doesn't want to eat. Brenda manages to get him to accept a tiny squirt. He swallows weakly and then droops in the nest, his eyes slits. "Please, please, please be okay," Brenda whispers.

"I think we need to be out of the room," she says.

We move to the dining room, a welcoming space with a fireplace and a view through sliding doors of the deck and seed and nectar feeders ablur with the wings of purple finches, titmice, orioles, towhees, grosbeaks, and hummingbirds. Inside, shelves and cabinets display treasures from the couple's life: rocks, fossils, antlers, crystals of red hematite and quartz, petrified wood, whalebone from the beach, an ancient anchor stone from an Indian canoe, an Audubon print of the water ouzel (also called the water thrush), and several of Brenda's smaller sculptures. They are abstract, with rough edges and holes, and look almost like they were found, not made. The holes let in light, which seems to change the sculpture; it's as if the holes let in time and movement as well. The forms remind me of kachina dolls, the movement of a walking ape, of beings part human, part animal. I ask her about her art.

Growing up in the mill town of Plymouth, New Hampshire, Brenda read an issue of *Life* on Matisse, and another on Picasso, and it changed her life. She dropped out of high school at sixteen and moved to Boston—"to find a mentor to apprentice to and be a sculptor." I looked at her incredulously. "That's what you did in the Renaissance," she said. "I didn't know any better." She found one immediately. "On a side street, near Harvard Square, beneath a bookstore, there he was, working in his studio. He was perfect! I went up two or three nights and pounded on the windows and he just yelled at me to go away. One night I yelled, 'I want to make sculpture and I want to be your apprentice!' And he yelled back, 'I teach at New

England School of Art and Design—if you want to work with me, enroll!'
I was devastated."

College seemed unlikely. Brenda's mother was the only one in the family to graduate from high school. Her father left school to join the navy in World War II and worked as a tool grinder and machine specialist at a knitting factory for twenty years—till it was bought by outsiders who closed it down. He lost his pension.

But Brenda did go to New England School of Art and Design and worked with the basement sculptor, Robin Benning. She funded her tuition with an inheritance left by her grandfather's sister—for a year and a half. Then came her first marriage, two daughters and a son, as well as the son of her husband's sister, whom the couple adopted when he was three—when Brenda was just about to go back to school again. But always she has found time to make her sculptures.

"Sculpture is about process, and this process is often fragmented. In a way, our lives are like that. A lot of people have trouble with transitions, with discontinuity. But this is what makes us grow. It's mysterious. You shift to a new platform and see things from a different perspective. Art does that. Wildlife does that. Wildlife and art reveal these transitions and demand we experience them. They—"

The timer sounds.

And at last, the smaller baby is hungry. Both birds eat with gusto, eyes bright. The littler one preens a wing feather with his bill—a balletic, careful motion, not a tortured one. "Very good!" says Brenda. And at last, the triumph: he rises in the nest and, clutching its softness with his feet, revs his wings. Brenda and I hug and wipe our eyes.

"You know that kind of awestruck, timeless feeling you get when you look at a great work of art?" Brenda says. "That sense of wonder, that sense of connection to something great and mysterious? It's the same feeling looking at a baby hummingbird."

Each is just a speck—a firefly, a flash, a brilliant atom. Yet each is an infinite mystery.

"It's a layer of this world we know very little about," Brenda says.

* * *

It's now that I propose to break the rule. "Let's name them."

Brenda concedes. Our interest in these babies is far more than professional. Frankly, we love them—not just because they are hummingbirds, but because they are *these* hummingbirds, distinct individuals. Besides, we're tired of calling them "the bigger one" and "the smaller one." Brenda asks me for suggestions.

I've been thinking about this. I suggest we name the older, stronger bird Aztec. The Aztecs admired hummingbirds so deeply that they adorned statues of Montezuma with their feathers. Aztecs believed that hummingbirds, forever chasing one another with sword-like beaks, were resurrected warriors, returned to life so they could continue their battles in the sky. Their word for the god of war, Huitzilopochtli, combines the Aztec words for "hummingbird" and "sorcerer who spits fire."

"Too violent!" says Brenda. "Something else."

Every culture to encounter hummingbirds has tried to name their magic. Ancient Mexicans called them *huitzitzil* and *ourbiri*—"rays of the sun" and "tresses of the day star." In the Dominican Republic, people call them *suma flor*—"buzzing flower." The Portuguese called them *beija-flor*, or "flower kisser." Even the scientists succumbed to hummingbirds' intoxicating mysteries: they classified them in an order called Apodiformes, which means "without feet"—for it was believed (incorrectly) for many years that a hummingbird had no need for feet. It was thought that no hummingbird ever perched, accounting in part for its sun-washed brilliance: as the comte de Buffon, Georges-Louis Leclerc, wrote in his 1775 *Histoire naturelle,* "The emerald, the ruby, and the topaz glitter in its garb, which is never soiled with the dust of the earth."

But for names we'll speak many times a day, I suggest two others. For the larger bird, Maya: the Mayans believed that hummingbirds were made from scraps left over from other birds and that their brilliant colors were a parting gift from the Sun God. For the smaller one, Zuni: like the Hopi and Pima, they believed that hummingbirds bring rain—which this part of California hasn't seen since January. Apt names, I feel, for both are named for blessings. And before long, Maya and Zuni will be embarking on the arduous journey to seize the greatest of all birds' blessings: the blessing of flight.

* * *

Bird flight is a confluence of miracles: Scales evolved into feathers. Marrow gave way to air. Jaws turned to horny, lightweight beaks bereft of teeth. (This is why many birds swallow stones: to grind their food since they can't chew it.) Hands grew into wingtips. Arms became airfoils. "The evolution of flight has honed avian anatomy into an extreme and remarkable adaptive configuration," anthropologist Pat Shipman writes in her wonderful book *Taking Wing: Archaeopteryx and the Evolution of Bird Flight.* But while most birds are made to fly, and the urge to fly is instinctual, flight itself must be carefully and painstakingly learned—a task of impressive complexity.

Consider the three basic methods of bird flight. The simplest, gliding flight, demands exquisite balance and judgment. Gliding flight exploits passive lift to counteract the pull of gravity. Vultures, hawks, and eagles ride the currents within thermals, rising columns of warm air. Albatrosses and petrels exploit different layers of wind speed above waves. Birds can glide for hours, expending very little energy.

Flapping flight—the way most birds fly—is more demanding still. Achieved by flexing wings at joints in wrist, elbow, and shoulder, it is powered by extraordinarily strong breast muscles. The wings move forward in a downward arc, propelling the bird forward and up. It is similar to the oarsman's power stroke or the action of a swimmer doing the butterfly. Movement then flows into the upward stroke, a recovery stroke, to start the process anew.

And finally, there is hovering, unique to hummingbirds. No other bird really hovers—kites, storm petrels, kestrels, and kingfishers appear to do so, but only hummingbirds can sustain this method of flying for more than a few moments. Instead of flapping the wings up and down, the wings move forward and backward in a figure eight. During the forward and back strokes, the wings make two turns of nearly one hundred and eighty degrees. The upstroke as well as downstroke require enormous strength; every stroke is a power stroke. Like insects and helicopters, hummingbirds can fly backward by slanting the angle of the wings; they can fly upside down by spreading the tail to lead the body into a backward somersault. Hovering becomes so natural to a hummingbird that a mother who wants to turn in her nest does it by lifting straight up into the air, twirling, then coming back down. A hummer can stay suspended in the air for up to an hour.

Hummingbirds are specially equipped to perform these feats. In most birds, 15 to 25 percent of the body is given over to flying muscles. In a hummingbird's body, flight muscles account for 35 percent. An enormous heart constitutes up to 2.5 percent of its body weight—the largest per body weight of all vertebrates. At rest, the hummingbird pumps blood at a rate fifteen times as fast as that of a resting ostrich, and that blood is exceptionally rich in oxygen-carrying hemoglobin. "There can be no doubt it reigns supreme over all the other birds of the world," writes Esther Tyrrell, "and truly deserves to be called the champion of flight."

While Maya and Zuni are still revving wings in the nest, I will fly home on a jet plane. By the time I return, they will be flying on their own—and nearly ready to fly free.

Years of experience have taught hummingbird rehabilitators how to ready baby birds to become champions of flight. No human can teach a hummingbird to fly, of course. Rather, Brenda thinks of herself at this stage as "a little gym instructor." The best she can do is provide the fledglings, at the right time, with the properly equipped gymnasium.

When an orphaned baby first leaves the nest and starts perching, it graduates to a larger basket, where it learns to fly from one perch to another. When that task is mastered, the bird moves to a laundry basket, where it begins to hover. All the baskets are encased in nylon mesh to prevent escape—and also to hold in the live fruit flies Brenda and Russ release there. Catching bugs is a further incentive to hone depth perception, strengthen muscles, and develop beak-eye and wing-tail coordination.

During my absence, Brenda e-mails me news of the babies' milestones:

July 1: Maya moving to #2 basket and we will introduce fruit flies. FUN. Zuni just fledged—not self-feeding. Zuni peeps for me to feed him.

July 3: Zuni is FINALLY self-feeding (yesterday) but still in #1 basket. Maya just moved to a #3 basket. I will move Zuni to a #2 basket tomorrow . . . transitions seem to be more challenging for Zuni.

July 6: Maya hovering and catching fruit flies in the #3 basket.

He is the perfect bird, right on track. Zuni is in the #2 basket, flying better, but still gapes in the a.m. I keep the two side by side outside by day and move them in at night. Moving Maya to small wire cage today.

The small wire cage is about two and a half feet by two feet by two feet. The bottom of the cage is covered with potted blooms, all shades of coral, pink, and crimson—penstemons, fuchsia, *Bravoa geminiflora*. The top half of the cage is uncrowded, with space for unobstructed flying. If you could hear the buzzing in there, you might conclude that this is a cage in which crazy people are keeping a bumblebee.

The bumblebee, I find when I return, is Zuni. I hear him long before I see him, and the sound, so loud and deep, makes me laugh. Its pitch is the mark of an amateur, Brenda tells me: "He hovers, but he's not good at it yet," she explains. "When he gets it, he'll make just a little buzz." Russ rushes outside to get him more bugs, so Zuni can demonstrate fruit-fly catching. The little bird zigzags after the flies, seizing them in a tweezer-like bill that has grown at least a quarter inch since I last saw him. Zuni's flights last only seconds and sound like a propeller plane—but to me, they are miraculous.

Next we visit the porch and the large white flight cage on the deck. About five feet tall, the cage once housed a pet parrot. But now it's been rechristened the Hummingbird Hotel. Furnished by Brenda's artist's eye, it's like a lovely garden inside, the floor carpeted with nine potted plants, the walls adorned with syringes, perches, feeders, and water for bathing. The cage is a beautiful setting for the stunning gem inside it: glimmering, sleek, green Maya. He zooms up and down, flitting from one bloom to another, hovering at exactly the right angle to extract nectar from the throat of each flower. His wings are a blur. It looks as if he is riding in a small tornado. "Isn't he great going up and down with his tail?" Brenda asks. "This is one of the things he had to perfect in this cage."

Maya has learned all he can learn from this largest of Brenda's cages. Tomorrow will be a momentous day: he will fly free and begin his life as a wild Allen's hummingbird.

* * *

In the morning, Brenda turns Maya's big cage so it faces the garden. We move Zuni's cage from the dining room table to next to the Hotel. When you move a cage, the hummingbird in it flies—as if Maya and Zuni both realize this is the proper way a hummer moves from place to place, rather than being carried.

The day warms. Zuni buzzes, rocking from side to side a bit as he flies. Maya flits from flower to flower, bathes, preens, then softly buzzes from one side of the cage to the other, as if pacing. Maya seems ready for the release. But I wonder: is Brenda? She has done this many, many times. This is, after all, the point of all those weeks of every-twenty-minute feedings, of curtailing her artwork and social life, of tears and prayers. How does she feel? "I feel confident I did everything I can," she says. "He can fly up and down. He can recognize the feeder, he knows many plants. This is what I want. I really want them to be wild birds." But every time she releases a hummingbird, her heart, too, flies out that cage door.

She summons Russ. "It's time."

Brenda draws aside the curtain of the screen and ties it back for Maya. He hovers, perches, then flies in zigzags from top to bottom as if the cage has grown too small for him. He seems eager for the wider world.

"Watch—there he goes!" cries Russ. But Maya is still buzzing inside the cage. He can't see the way out. He sits on the lower perch in the middle of the cage, as if considering his options.

"Sometimes," Russ tells me, "they fly out fast and—*bip!*—they're gone." Other times, they just sit awhile, understandably hesitant to leave the protection of the cage. And sometimes, Russ said, they fly out, and then back in, before they finally take off.

Maya feeds from the flowers at the bottom of the cage. He flits to his feeder. And now—at 11:01—he darts out of the cage and dives down a slope into a bay tree below the house, a male Anna's in hot pursuit. It's a flight easily in excess of sixty feet—twenty times farther than he has ever flown before.

Zuni peeps piteously in his flight cage. In the wake of his sibling's achievement, "he wants his food, he wants his mother," says Brenda. Though there are feeders he can reach easily in the flight cage, she feeds him by hand. "I'm worried about him. I've been worried about him ever since we got him."

We scan the yard with binoculars. Released Anna's always vanish, but Allen's are different. They're likely to stick around for a few days, so we hope we'll see Maya again. We want to document his first days of freedom. Brenda leaves the Hotel door open. She wants to test her hypothesis that newly released Allen's welcome the chance to roost there in safe familiarity for their first few nights of freedom.

But Maya has seemingly vanished.

11:30 a.m.: We check for Maya on a circuit every thirty minutes. We first peer from the garage in the direction of the Hummingbird Hotel. We next check the feeder by the studio. Two male Anna's shoot by like flaming comets. A third hovers at the door of the vacant Hotel, eyeing the flowers and feeders inside. Finally we walk through the hummingbird garden. The feeders and flowers buzz with Anna's of both sexes. But no Maya.

12:00: Two Anna's zip through the garden, fighting. A newly fledged Anna's tries to feed by the studio but is chased away, again and again, by adult males, flashing their rose gorgets. The fledgling huddles, fluffy and forlorn, in the branches of an oak. But where is Maya?

"You wonder," says Brenda. "It's a big world . . ."

"He'll come back," says Russ gently.

"Sometimes he would flag and sputter like a little helicopter in trouble . . ."

"He was flying fine," Russ answers. "He'll be fine."

12:30 p.m.: We watch with binoculars from Brenda's studio. The world is crowded with hummingbirds. Anna's zoom everywhere: at the corner feeder, at the lily of the Nile, at the red salvias. "I don't see anything," says Brenda. "Where is my baby?"

What scares us most is we have not seen him feeding. To survive, a hummingbird must consume the greatest amount of food by body weight of any vertebrate. A film Brenda loaned me claimed that a person as active as a hummingbird would need 155,000 calories a day—and the human's body temperature would rise to 700 degrees Fahrenheit and ignite! To fuel the furious pace of its life—even resting, it breathes 250 times a minute, and its heart pounds at five hundred beats per minute—a hummer must daily visit fifteen hundred flowers and eat six hundred to seven hundred insects. If the nectar alone were converted to its human equivalent, that would be fifteen gallons a day.

Food is so precious to hummingbirds that they defend "their" flowers and feeders against all comers. They sometimes even chase away hawks and crows. Their main rivals, of course, for the food are other hummingbirds, and hummers' belligerence toward their fellows is legendary. Russ recalled the title of one article about them headlined "So Little. So Pretty. So Mean." In their book *Hummingbirds: Their Life and Behavior,* Ester Quesada Tyrrell and her photographer husband, Robert, document fighting hummingbirds trying to stab each other's eyes out with their bills. The couple has seen a hummer grab the rival's bill in flight and, locked in midair, the two birds fall to the ground together—then rise to continue the battle. A male humming-bird may spend one minute in fifteen lapping from a favored feeder—and the other fourteen defending it. But the way hummers usually kill rivals is bloodless. They simply chase the rival bird from food until it runs out of energy. It can enter a state called torpor in which the body temperature, normally more than 105 degrees Fahrenheit, falls to below 70. Torpor can also be caused by chill, and it can take an hour for a hummingbird to rouse from it; during this time, even the most slow-moving predator—even a possum—can take a hummingbird. But if a predator doesn't get it, starva-tion will. A torpid hummingbird may perch next to a nectar-laden flower but be too weak to summon the energy to drink from it.

Is Maya now torpid, frozen with fear and exhaustion? Normally a fledg-ling hummingbird would call its mother to feed it. But we can neither hear nor see our baby. If he's in trouble, we are impotent to help. I feel a growing lump in my throat. Please please please, my heart pounds. Just a glimpse. Just one more glimpse.

It's as if Brenda can hear my thoughts. "But then, the more you don't see them, the better," she says. "We really do want them to leave."

1:00 p.m.: A female Anna's hovers, lapping from the purple globular blossoms of a lily of the Nile. A male flashes his gorget as he feeds from a potted fuchsia. A baby perches high in the madrone tree, cleaning its bill on a branch. But it's an Anna's, not our Maya.

1:30 p.m.: Only yellow jackets at the feeder by the studio. An Anna's perches on the lowest branches of the oak. Maya is nowhere to be seen.

1:40 p.m.: Was that him? While Brenda is in the kitchen, I think I see him trying to drink from the studio feeder. But the youngster is chased immediately by a male Anna's with a flashing magenta gorget. A baby Anna's eyes the studio feeder, decides on bugs instead, but is chased away anyway. Another male flashes his deep rose gorget below from the laurel. Was I imagining Maya?

1:53 p.m.: No—it *was* him! As I walk from the studio toward the garage, I find him perched on the wire fence surrounding the deck. He sips from a potted fuchsia, then perches on the wire behind the flowers, trying to hide from the Anna's—but they chase him off instantly, careening after him like bullets. He buzzes my face, hovering in front of me, as if he is asking me for help. I rush in to get Brenda.

When we come out, he's perched on the wire again, fluffed up and peeping like his life depended on it—and it might. Maya weaves his head back and forth like a desperate child, as if pleading: Help me!

"He looks really stressed out," says Brenda. Just as a mother hummer would do, Brenda rushes to feed him. She holds up one of the extra feeders she's put out along the fence. Maya hovers before it—drinking, at last, unmolested. He perches, then hovers for a second drink. This helps. He stops peeping. He preens. He cleans his bill with his foot, then wipes his beak on the wire.

"It's a rough, tough world out there," says Brenda. "All those Anna's attacking! These first couple days are so critical. Between the other hummers and all the predators out there . . . but he has to do it . . ."

Brenda brings more nectar feeders and sets them along the fence. Attached to the fence with a wad of putty, a syringe can become a hummingbird feeder in a jiffy. Brenda has an almost endless supply. The idea is to overwhelm the Anna's with abundance so little Maya gets a chance to feed. People who set out feeders and plant wisely for hummers know this well. It's possible to have as many as forty of them in your yard at once—even though every bird thinks every feeder and flower belongs to him alone.

Despite the abundance, the Anna's continue to chase. But Maya is often able to steal a drink before he flies off. He's learning, and he's clever. At 2:48, he flies to the studio feeder and is chased away by the Anna's. We're impressed with the sophistication of his escape: he maneuvers his tail to whip and loop down toward the laurel below and around the poison oak to safety. At 3:50 he perches on the wire and is challenged bill to bill by a hovering male Anna's—and holds his ground. Sometimes he uses one of us as a human shield. At 4:10, he reappears on the wire near the fuchsia when he is dive-bombed by an Anna's. Both fly toward Brenda, who is watering the hummingbird garden. She offers the spray to the Anna's, who takes a bath— and now, with feathers soaked, will need to dry and preen, giving Maya a chance at the feeders.

"You're getting it, baby!" Brenda cries. We hug and high-five. From the kitchen, Brenda brings sparkling water in long-stemmed glasses. "A toast to our little baby!" she proposes.

But it could be too early to celebrate. At 5:20, Maya reappears, feeding on fuchsia flowers by the garage door. In a flash he vanishes, shooting over the roof with a speed and purpose we had never before seen. No other Anna's are in sight. In fact, even the seed feeders are curiously vacant. Then—BAM! Something hard and heavy hits the wood railing of the deck, just yards from where Maya had perched. It's the steely talons of a sharpshinned hawk.

Twenty-nine minutes later, Maya returns. He perches on an unused bird cage near some potted fuchsias. As two Anna's fly overhead, he appears to flinch, but doesn't flee. When a turkey vulture swoops close to the porch, I watch carefully to see what Maya will do. Humans often confuse vultures with hawks, as both are big birds who soar. Many field guides class vultures with raptors, even though hawks are hunters and vultures eat carrion. But Maya doesn't make the same mistake. He watches the vulture with interest but does not shrink from it or fly away.

How does a five-week-old hummingbird know the difference between a deadly hawk and a harmless vulture? The knowledge could be innate. Or it could be the result of careful observation. Hummingbirds are curious and astute observers. Backyard hummers quickly learn to recognize individual humans and approach people who feed them—even if they are not carrying food. (Sometimes they'll hover at windows to attract a particular person to come out and refill an empty feeder.) In *Hummingbird Gardens*, authors

Nancy Newfield and Barbara Nielsen write about a Costa's hummingbird whose nest fell apart in an exhibit at the Arizona-Sonora Desert Museum. While an assistant constructed an artificial nest, keeper Karen Krebbs held the five-day-old nestlings. Though wary of the assistant, the mother bird clearly recognized and trusted the keeper—so much that she perched on the woman's hand as she fed her babies. Hummingbirds remember what they see. Many folks who feed hummers report that if they're slow to get their feeders out in the spring, hummers show up and hover right where the feeder used to hang the year before.

Maya's quick flight from the hawk, and his worldly nonchalance with the vulture, give me some reassurance. Maya is a competent little bird. But he's still a baby. And night is coming, with its cool temperatures, hunting owls, and roaming cats. I'm glad Brenda's leaving the door open to the Hummingbird Hotel.

The release cage is a safe haven. Once, two wild Anna's took refuge there during a winter storm. One chose a perch near the top; the other chose the farthest possible spot away from him. Normally hummingbirds can't stand the sight of another hummer who's not a potential mate or its own young. They eyed each other warily but spent the night together nonetheless.

At 7:20 p.m. we find Maya perched serenely on a stem in the release cage. Through the kitchen window, we watch him till dark, until he is just a silhouette that dissolves into the soft, moonlit night.

Having spent the night, happy and comfortable, inside the release cage, by 6:30 a.m. Maya is again a blur: flying in and out of the Hotel, sipping from feeders inside and out, inspecting flowers in the garden. Zuni, too, spent the night outdoors—his first—in the small wire cage right next to the Hotel. "He seems very happy," says Brenda sunnily. "I think they like feeling the cool air."

Today Brenda plans to move Zuni into the Hummingbird Hotel. Since he hatched two days after Maya, that would put Zuni just about on schedule for a release three days from now. Typically birds spend three days perfecting their hovering in the Hotel before release. But Zuni is not a typical hummingbird. "He may panic. Every time we've moved him up a cage, he's panicked," Brenda says.

Once the day has warmed, Brenda captures Zuni in her cupped hands while I hold aside the screening by the door to the Hummingbird Hotel. Zuni perches on an upper right stem. He surveys his new surroundings like a first-time visitor to a great cathedral: he seems awed as he moves his head, bird-jerkily, in every direction above him. But within ten seconds, he hovers to a fuchsia in a top-level vase. Russ releases some bugs in the cage and Zuni flits about snapping them out of the air.

It's a delight watching both babies expand their world. Midmorning, Maya hovers in front of one of Brenda's statues—an armless torso, about three feet high. There seems little hope of nectar here—it's not red, a color many flowers sport specifically to attract hummingbird pollinators, as it's a color bees don't see. The statue is white concrete. There is nothing flowery about it. Maya simply seems curious. He spends nearly a minute examining it—more time than many people invest in any given piece of art at a gallery. Zuni, meanwhile, is clearly practicing his flights and hovers. In the large cage he can cover nearly three times the airspace of the smaller one. He zooms up, down, back, and forth. The stems of flowers bounce when he lands on them. It reminds me of a child playing on a trampoline.

Hummingbirds do seem to play, and to enjoy it. Authors Newfield and Nielsen recount how one Anna's flitted through the spray of a fountain in a California garden: "One day she discovered that she could ride the stream, a solid jet of water about three quarters of an inch thick. Flying at right angles, she alighted on the jet, as though it were a branch, and permitted it to carry her forward. Over and over she did this, apparently enjoying the stunt. She seemed to be playing rather than bathing." A hummer will also bathe on leaves slick with dew or rain. It will slide down the leaf, moistening its breast and shaking its feathers while still in motion. It must feel lovely.

The following morning, we can see how much Zuni's hover has improved. An adult Allen's hovers with an upright posture, holding the tail high with little movement; when moving up and down, it flicks the tail. Up to now, Zuni's posture has been hunched, his tail wavering. "He's getting it!" Brenda cries.

But when we check on him again at 8 a.m., the little bird who was doing so well is now a pathetic, peeping puffball.

He issues a peep every second, one after another. Each sounds like it could be his last. When Brenda approaches, he gapes as if he wants her to feed him. But when she holds the syringe to his beak, he won't stop gaping long enough to drink. Finally he inserts the tip of his beak into the feeder and we see his throat flutter as he flicks his tongue.

What has gone wrong? We back away from the cage and watch from the kitchen window. And now the problem is clear: Anna's are challenging him from outside the cage. They buzz by, hovering in front of the flowers inside, eyeing them jealously. They discovered the place was full of flowers and feeders while the cage door was open yesterday. The males are desperate for food: though the species is resident in the Bay Area year-round, many individuals retreat from the northern edge of their breeding area in the fall. The males are first to leave, and they must increase their body weight by half for the trip. "It's such an intense time for them," says Brenda, "and there are so many baby Anna's!" To Zuni, it must feel like his house is surrounded by thugs, peering through his windows, eyeing his stuff.

Zuni is right to be afraid. "An Anna's will kill an Allen's," Brenda says softly. Anna's will occasionally stab with their bills but more often they kick their opponents midair as they hover. They taunt each other with a tail gesture that mimics this action. "Instead of giving him the finger, they're giving him the tail," I say. One brash Anna's even dares to perch on the cage bars and stick his sword-like bill inside. Zuni dissolves into peeping, like a child bursting into tears.

"If he's like this inside the cage . . . ," Brenda worries. "He's got plenty of bugs, and more flowers than he'll ever need . . ."

"What can we do?" I ask.

"We might have to move him back to the other wire cage," Brenda says.

"Oh no! That's worse than being left back!"

Brenda thinks for a moment. "He's safe—but he doesn't know it," she says. She slides a glass panel into one side of the cage, to repel the Anna's thrusting bills, and covers another side with a blanket. At least Zuni won't have to look at frightening Anna's bullying him from all sides. She also moves the cage. It's currently positioned, she realizes, directly in the flow of traffic between two of her most popular hummingbird feeders—the one at the kitchen window and the one outside her studio. Just a few yards away,

Zuni's cage won't be blocking traffic anymore. And finally, she decides we should go run an errand in town. Perhaps if he can't see us, he won't focus on calling for his mother.

But when we return, Zuni is so despondent he won't even fly to the syringe when Brenda holds it up to him. He's like a tot who has cried for so long he's forgotten what he's crying about—and now is crying because he's exhausted from crying. Brenda has raised three children and more than eighty baby hummingbirds. She knows what to do. Patiently, she holds the syringe to his beak while he perches. Finally he summons the wit to drink.

The food revives him, but soon he's peeping again. Brenda spritzes him with water. He's forced to preen, giving him something to do other than perch frozen in terror. He fluffs, runs his beak through his feathers, wipes his beak on the perch. Dry again, he zooms around the cage as if nothing had happened, but thirty minutes later, he's puffed and peeping again.

"You know how kids get cranky and nothing can console them?" Brenda muses. "I think the Anna's just messed up Zuni's whole day. He's got plenty of food. He's flying well. He can do it—maybe he'll be all right tomorrow."

Zuni improves each day. On Saturday, an Anna's sticks its bill into the Hotel right in Zuni's face—and Zuni holds his ground. When the Anna's leaves, Zuni hovers for two seconds—an aerial victory dance?—then perches on a high stem. Nary a peep. Now he understands the cage is a safe barrier.

Maya buzzes by from time to time, but not as often. He's enlarging his territory. Brenda spots him over in the neighbor's yard. "He was flying so fast you would never see him if he didn't stop!" she tells me. While Maya is drinking from Brenda's studio feeder, a juvenile Anna's zooms in. Like a targeted missile, Maya rockets after him, chasing him from the feeder and over the roof.

Sunday dawns cool and windy. Brenda had hoped to release Zuni today, but the wind is strong enough to lift a tablecloth off the picnic table on the deck. And to our horror, Zuni's right wing is fluffed out at the shoulder. When he flies, he lists to the right. Did he fly up the side of the cage and hit a perch? Did he crash into something while frightened by an Anna's? Was he buffeted by a gust of wind? Whatever it was, it happened while he was safe in the cage. Brenda is exasperated. "This bird has been one problem after another!" she says. Today's problem seems to be a single, twisted feather. Brenda spritzes Zuni with a mister, hoping he'll set the feather right when he preens. Zuni, always an enthusiastic bather, sets about the task rather cheerfully. He's not puffy or peeping; he doesn't seem fearful or in pain. The twisted feather isn't bothering him as much as the setback is bothering us. "He'll be releasable," Brenda promises. "I just can't take the chance to release him today."

But we will celebrate anyway. Brenda has arranged for me to do a reading at a local bookstore, and afterward she and Russ are hosting a hummingbird garden party. Russ's parents and brothers, neighbors, and friends from WildCare are invited; a friend of mine is driving all the way up from Fresno. Also on the guest list are two people we have never met: the couple who first found Zuni and Maya. We've always wondered about them.

The original rescuers are the last guests to arrive, after most of the others have left. Michelle Earnhardt, thirty-nine, pretty and petite, has long blond hair and is wearing white pants, a long blue scarf, and a leather jacket. Her heavyset husband, Matt, thirty-seven, towers over her five-foot frame like a big bear in shorts and a T-shirt. She's a hairdresser; he's a contractor. They wanted to get here earlier, Michelle explains, but the drive from Novato, normally half an hour away, was clogged with traffic. It's clear that coming here was very important to both of them.

Michelle tells us how they found the twins. They had just returned from their wedding and honeymoon in Hawaii when they first saw a hummingbird frequenting the passionfruit vine on the trellised entryway to the side yard. They were delighted. It was just six feet from the front door. "We saw a hummingbird there all the time," said Michelle. "And then, one afternoon, it wasn't there."

Hours passed, and though the couple watched and waited, still they didn't see the hummingbird. Both were nearly sick with worry. They both love animals. Before they moved to Novato, they had rented a house in Mill Valley that came with a single goldfish who lived in a barrel. What was he eating? they wondered. He must be starving! So they named him Starvin'

Garvin and fed him garden worms. When they moved to Novato, they left Garvin behind—after all, he didn't belong to them. But that first night, Matt couldn't sleep for worrying about the fish. He drove back to Mill Valley in the middle of the night and returned with Garvin and his barrel.

To Michelle, hummingbirds were especially important. She had once gone to a sort of fortune-teller, a shaman, who told her she had a totem animal, a creature who was constantly with her, guiding her, even though she couldn't see it. Her totem was a hummingbird, the shaman said, and it stayed right in front of her forehead. Michelle was not surprised to learn this, she said, for hummingbirds have appeared to her and guided her many times at important junctures in her life.

When she was a teen, her mother and later her friends had taken to calling Michelle a hummingbird, since she was so petite and quick. Michelle had spent part of her childhood in Iowa, and even though the family now lived close to one another in California, at one point Michelle had contemplated moving back. Her mother was thinking about this when one day, the song "Hummingbird" came on the radio, with its lyrics "Don't fly away." Immediately afterward, a hummingbird hovered in front of her face. Michelle's mother phoned her daughter: "You can't move. I got a sign." Michelle is glad she listened. If she had moved, she never would have met Matt.

A hummingbird helped her right before her wedding, too, she told us. Just weeks before the wedding date, she, Matt, and ten other people flying to Hawaii for the ceremony lost their tickets due to airline problems. "I was so upset," she said. And then one cool morning, Matt found a hummingbird on the back porch, unmoving. He picked it up and it flew away. "And I thought," said Michelle, "if a hummingbird can stay still that long, I can, too." Suddenly, she felt at peace.

So Michelle and Matt were determined to try to find the missing mother hummingbird and help if she was in trouble. She and Matt leaned a big ladder up against the side of the house so he could get a better view. That's when he spotted the nest, partially hidden in the passionfruit vine. He parted the leaves. Two pink, naked babies popped their heads up, gaping voicelessly for food.

Michelle called WildCare, but the rehab center was closed. The couple didn't know what to do.

That night before they went to bed, Michelle climbed the ladder to check on the babies. They weren't moving. She wept all night.

In the morning, before Matt left for work, he climbed the ladder to check the nest one last time. He brushed aside the leaves, and the larger baby popped open its mouth! The other baby seemed barely alive. But there was hope.

It was 7:00 a.m. WildCare wouldn't open till 9:00. So Matt drove the twenty miles to his contracting business in Tiburon, organized his work crews for his absence, then drove back to Novato. He ascended the ladder and clipped the nest free of the passionfruit vine. He placed the nest in a shoebox and, with the babies periodically gaping weakly in the passenger seat of his pickup, drove the fifteen miles to WildCare.

"Want to see them?" Brenda asks. The five of us—Brenda and Russ, Matt and Michelle, and me—head out to the deck. It feels like a family reunion of sorts, for all of us had a hand in Zuni and Maya's survival. Maya zips by like a comet. "That was one of them," says Brenda. Now they approach Zuni in his cage for a longer look. "Oh!" Michelle whispers, and touches Matt's strong arm with her fingertips. When she last saw this little bird, he was pink, naked, all but dead. Now he is a feathered sequin, hovering before a flower.

"He gets himself in trouble," Brenda tells Michelle. She speaks about Zuni with the fondness of a teacher talking to a mother about a favorite but difficult child. "Somehow he got off-kilter today. But he's okay now. His chances are really good."

In three days, Zuni will be released; two days later, both he and Maya will vanish, headed on the fall migration.

We don't know that yet as we stand before the tiny bird in the release cage. But already, we can imagine him flying to Mexico.

Birds Are Fierce

M y introduction to falconry is a bloody one.

On a cool, grey mid-October day, master falconer Nancy Cowan, a petite, blue-eyed blonde in her late fifties, hands me the most beautiful bird in the world: a four-year-old female Harris's hawk named Jazz. A deep, coppery brown, with reddish shoulders and a white tail tip, Jazz stands more than twenty inches tall, weighs thirty ounces, and her outstretched wings span nearly four feet. Her profile is regal and knowing. Harris's hawks are big, but their appetites are bigger. In their native cactus deserts in the American Southwest, they hunt moorhens as big as ducks and jackrabbits that can weigh more than twice as much as they do.

Nancy has offered me a choice: of the dozen or so birds of prey she and her falconer husband keep on their rural New Hampshire property, I could work with Jazz or with Emma, the lanner falcon. Emma is also beautiful. With a slate back and wings, creamy belly, and chestnut crown, her kind is the species pictured in Egyptian hieroglyphics. But Emma is smaller than Jazz by a third. And, Nancy explains, Emma has been raised by humans. At the mature age of sixteen or seventeen years old, Emma is exceptionally docile and calm.

Jazz is not. Nancy had warned me: Jazz is "feisty," sometimes uncooperative, and she "doesn't like hats." (How Jazz expresses displeasure is left unspoken—but looking at her curved obsidian talons and sharp black beak, I am glad that despite the cold I have come bareheaded.)

But I am not afraid. It's her wildness I want from the moment I set eyes on her.

Stepping from Nancy's falconry glove to the one loaned to me, Jazz's huge yellow feet grip my left hand with shocking strength. It is wise to be sheathed in leather. Otherwise, simply by perching, her talons would rip the skin of my hand and wrist and could easily pierce me to the bone. I am impressed by Jazz's feet, but I am awestruck by her huge, mahogany eyes. They look past my face, past my soul, as impassive and hungry as fire. Her eyes seem to be devouring the world.

I know I don't matter to her at all. Yet to me, she is everything.

Why do I love her so immediately? I love that she is big; I love that she is fierce; and I love, too, that she might be unpredictable. She is essence of hawk, a bird so unlike anyone I have ever personally known. And here she is on my arm.

I pretend that claiming Jazz isn't greedy. That leaves the smaller, calmer bird, Emma, for my friend Selinda. It didn't take much to persuade Selinda to accompany me; she loves animals as well as a good adventure (her first job out of college was working as a geologist in Alaska). But attending the half-day introductory course at Nancy's New Hampshire School of Falconry was my idea, and I reckon if one of us might get hurt today, it should be me.

Our instructor chooses to work with Banshee, a teenage peregrine falcon. The birds of prey recently reintroduced to cities to help control pigeon and starling populations, peregrines dive through the sky after their prey at a heart-stopping two hundred miles an hour. Her head is capped in black, her back and wings a deep, shiny blue, like the skin on a blue shark. Five inches shorter than my big Jazz, she seems tiny, precise, a knight-errant clad not in chain mail but in feathers.

We start walking down Nancy's gravel driveway, amazed that these majestic, predatory birds sit sedately on our fists.

Then Banshee bites Nancy in the face.

The attack comes without warning. It's a hell of a bite. Later Nancy explains to us that being bitten by a peregrine feels like being punctured by a staple gun. A notch in the bird's curved bill, an adaptation for crunching the vertebrae of its prey, makes the bite particularly messy and painful.

Blood gushes from the wound. What will Jazz and Emma do? I worry they might attack at the sight of blood, as my chickens did, but they ignore it.

Selinda and I, however, gasp in distress. "Don't worry about it," says Nancy. "Banshee's a teenager. She's just being a brat." Though I periodically dab at the wound with a tissue, the blood flows down our instructor's cheek and drips off her chin for half an hour.

Nancy is used to this. And so apparently are the neighbors. Drivers slow their cars and wave genially. Only Selinda and I seem to think anything of the sight of three women walking down a country road with birds of prey perched inches from their faces, one of them dripping blood onto the road. Nancy and her falconer husband, Jim, have lived here for twelve years. Everyone knows about their birds.

Only one vehicle stops. The driver greets Nancy (without a word about her wound) with the news "I've got something for you." He reaches into the backseat and pulls out a dead woodcock—a medium-sized, sandpiper-like bird of New England's fields and meadows—and hands her the corpse out the car window. Food for the hawks, I assume. But later, I learn this would be her dinner. Clearly, I have entered a strange new world. Selinda and I— vegetarians who mourn roadkills and weep over books in which animals are hurt—are taking this course because of our love of birds. We never thought that, less than thirty minutes into the course, we'd be facing violence, blood, and death. But what shocks me more is this: though I'm sorry that Nancy has been bitten and I'm distressed that the woodcock has been killed, some-how, in the presence of these birds, blood and death are not repulsive. I feel myself being drawn to a mind wholly unlike my own.

* * *

"God, birds are mean!" my friend Mike Meads once remarked to me. Mike had once been a farmer and had personally lopped the testicles off thousands of sheep; who was he to talk about meanness? Later, he had become a well-known New Zealand entomologist. I noted that insects were not known for practicing loving-kindness. But many others have remarked about this aspect of birds.

Even a pet parrot, who tells you he loves you in English, who showers with you each morning, and who shares your dinner at night, possesses a sort of ferocity. Living with a bird, as animal behaviorist Rebecca Fox told documentary filmmaker Mira Tweti, is not like living with a feathered child. "You should really think of it as living with an alien," she said. For even caged birds that have been kept as pets for many centuries remain fundamentally wild.

Unlike domesticated mammals such as dogs and horses, who are generally grateful for food and affection, your bird may not like you. Your parrot, for instance, might get mad at you—for something like going on vacation or away to college—and stay mad for the rest of its long life. Your pet parrot won't hesitate to punish you, sometimes ruthlessly.

But this isn't meanness; it is something else. Some would call it savagery. I call it wildness.

Birds are like us in so many ways that sometimes we forget we are from widely separate lineages. Birds are wild in a way that we don't experience in our relationships with our fellow mammals. And nothing, I found, brings one closer to the pure wildness of birds than working with a hawk.

"People have such misconceptions about these birds," Nancy tells us. Surrounded by framed prints depicting scenes of hawking in medieval Europe, Selinda and I sit in the warmth of the tasteful dining room of Nancy and Jim's 1789 house for the lecture portion of the falconry course.

"Some people think, 'Nancy owns this bird, so it's a pet,'" Nancy says.

But Nancy's relationship to her birds is utterly different from that of a person to her pets. "They might live with me, I might feed them," she tells us, "but they are wild. These are predators. And that is the beauty of them."

Raptors' wild beauty feeds yet another popular misconception. Folks are drawn to falconry for all sorts of wrong reasons, Nancy says. And one of them is this: "People see the bird as an ornament—they are thinking how cool they will look with one on their arm."

Selinda with Emma the lanner falcon

One can see how a person might succumb to that allure. Falconry, "the sport of kings," connects its practitioner with a romantic, proud history, stretching back to ancient China, India, Egypt, Persia, and Babylon, thousands of years before the existence of Rome. At one time, the type of falcon an Englishman was allowed to own marked his rank: a king carried the gyrfalcon; an earl, the peregrine; a yeoman, the goshawk; the priest, the sparrow hawk; and a servant, the kestrel. It is a sport with its own battery of accoutrements, including beautifully tooled leather gloves for the falconer and elaborate, often feathered hoods for the birds, sometimes considered works of art.

Falconry also has a language all its own, known only by the shared brotherhood of fellow falconers. Some of the words are needed to describe the sport's many accessories: "jesses," attached to leather anklets around the bird's legs, are soft leather loops to which one can hook a length of rope or a tether to the falconer's glove. The "bewit" is a slip of leather attaching bells to the feet, so you know where your falcon is. The "creance" is the long, light cord for tethering a falcon in training.

Falconry hoods, from the collection of Jim Cowan

Special words describe the activities unique to training and caring for a bird of prey. "Imping" is the act of mending a broken feather. "Manning" describes training the young bird to be carried on the fist. "Seeling" is the word for the ancient, now-abandoned practice of sewing the bird's eyelids shut—temporarily deprived of sight, the bird is rendered dependent on the falconer and more easily trained.

But much of falconry's secret language underscores, like a promise repeated again and again, how special these birds are, how different from all other beings. Though many bird species hunt, kill, and eat other animals— from the shrike, a songbird also known as the butcherbird, who kills and then uses thorns to skewer other birds to store them prominently for a later meal and attract a mate, to the worm-eating robin—birds of prey are exclusively predatory. They are also known as raptors (daytime raptors, more accurately, to distinguish them from the unrelated and mostly night-loving but equally predatory owls). Sometimes they all are just called hawks. They include some three hundred species that go by various names: hawks, eagles, falcons, harriers, kestrels, kites. (And to make it more confusing, the English use different words than we do; for instance, their buzzards aren't our vultures, but what we would call our red-tailed and ferruginous hawks.) They live all over the world. They are the tigers of the air. They hunt like no other predator.

The language of falconry honors this difference. The falcon isn't sleeping, like ordinary birds or mammals; it's "jonking." When it cleans its beak and feet after eating, it's "feaking." The act of hiding the food with outspread wings and tail while it eats is called "mantling." A bird of prey, in fact, is so rarefied that it doesn't even shit like the rest of us. Hawks "slice"; falcons "mute."

My interest in falconry, though, isn't in its language or tradition. From falconry I want only one thing: to get closer to birds of prey. Majestic, graceful, strong, big, brave, and smart: who would not hunger for such company?

Of course one can watch hawks at a distance—and I do. I'm always looking for them and frequently find them. They are more numerous than you think. Along any highway, you can sometimes count dozens, perched singly on trees near the road, looking for small animals among the mown grass. In the fall, from certain mountaintops, you can see hundreds, sometimes thousands of normally solitary hawks in a day, migrating en masse to southern wintering grounds. In the summer, if you know where to look, the huge stick nests of certain birds of prey are easy to find. My husband and I are blessed with a pair of bald eagles who nest near a lake not far from our house. And one can also watch birds of prey in captivity, at zoos, nature centers, and attractions like Sea World that give falconry flight demonstrations.

But I long for a more intimate connection, to see, close at hand, something of what it is like to be a bird of prey, to try to understand what is in their heads. As predators, what do they show us about other birds? I wanted to touch these birds' fine, ancient wildness, this pure savagery bereft of evil. And there is no other way to begin to do this, except through falconry.

The writer Thomas McGuane calls falconry "one of man's oldest and most mysterious alliances in the natural world." It is the art of partnering with a bird of prey. And as Nancy says, "You are not the master. You are the junior partner." If your hawk doesn't consider you a good enough partner, the bird will fly away. At any time, a bird may leave the falconer—for days, or weeks, or forever. But I have also read of birds who so value their partnership with their human that they stick around even when they are not handled or confined. In Stephen Bodio's splendid *A Rage for Falcons,* he writes of hawks who don't travel to the hunting grounds confined in carriers, but fly into the car for the trip, fly out to hunt, and fly back in to go home. Steve knows a goshawk who lives loose in the woods but flies to the falconer's whistle to join him for a hunt. He knows a pair of wild goshawks who come to another falconer's hand.

Few people understand the true nature of falconry, Nancy tells us. Oddly, bird-watchers often look down on falconry; some consider it a form of slavery. Others dislike the birds themselves, for raptors not infrequently attack and kill birds at feeders. In the United Kingdom, rogue members of the pigeon racing and game shooting communities have been known to intentionally poison, trap, and shoot hawks, particularly peregrines, a crime that occurs so frequently that it could threaten the falcon's survival. In the United States, a fourteen-month investigation uncovered that thousands of peregrines and hawks in Oregon, Washington, and California were being killed by members of pigeon clubs who specialize in Birmingham rollers—a type of pigeon bred for a genetic anomaly that triggers a seizure midflight, sending the bird spiraling downward until it recovers before hitting the ground. Naturally, no raptor can resist such an easy and obvious target. But the breeders who perpetuate the defect persecute the predators by trapping, shooting, poisoning, clubbing, and, in some cases, torturing the raptors, including dousing the birds with Drāno. Perpetrators convicted in 2009 received no jail time, prompting public demand for tougher penalties.

But even those who admire hawks and falconry often misunderstand them. "Some people come to me with a mystical outlook," Nancy says. "They think, 'Oh, this bird loves me.' They think they have some kind of spiritual relationship with the bird." This view, she tells us, is as misguided as considering the hawk a pet or an ornament—and in its different way, just as demeaning. "I can't imagine anything crueler to do to a living being," she says, "than to try to make it into something it's not.

"The only people who understand birds of prey," she says, "are people who have worked with them for a long time. And every time I work with them I learn something new. You accept that your life will be changed by being a falconer."

Though I do not know this yet, Nancy is right: my life will change, too. Over the course of our long, continuing association, Nancy and her birds will show me a kind of relationship I had never known was possible with any living being.

* * *

The saker falcon in the hallway looks like a sculpture. The bird sits on its perch immobile, its head covered with a leather hood. Hoods have replaced the old practice of seeling the eyelids of a hawk in training. You would think that being blinded would send a bird into a panic, but no; a hooded bird doesn't thrash or struggle; it won't bite you or grab you in its talons. Instinctively, the bird knows better. It stays so calm that you might not even know that a hooded bird is alive. And in a sense, it is not. Putting a hood on a bird is like extinguishing a candle.

But then Nancy performs a breathtaking act of magic. She pulls off the hood. Instantly, the inert sculpture comes vividly alive. The flame of his soul leaps to life. Brown with a pale head, he stands tall and alert as an officer at attention: his name is Sabretache, for an accoutrement of the British military uniform that hangs from the saber belt. The intensity of his gaze fills the room.

"These eyes are carrying huge messages," Nancy tells us. One of the defining characteristics of birds is the crucial role and astonishing acuity of their vision. Flight, after all, demands excellent eyesight. For birds who hunt on the wing, the eyes are developed to an extraordinary degree: an eagle riding a thermal at one thousand feet can spot prey across a distance of nearly three square miles. In this way, raptors are superbirds. They have developed the avian sense of sight to perfection. A raptor's vision is the sharpest of all living creatures.

All birds' eyes are huge in proportion to their bodies. A person's eyes take up only 2 percent of the face; a European starling's account for 15. A great horned owl's eyes are so enormous relative to its head that if human eyes were comparable, they would be the size of oranges. Their eyes are so important to birds that, like various reptiles, sharks, and amphibians, birds have a transparent or translucent third eyelid, the nictitating membrane, to protect and moisten the eyes while retaining visibility. Vision literally sculpts birds' every movement: one reason that birds seem to move in such a jerky manner, as cassowary expert Andy Mack explained to me, is that the bird is actually keeping its head remarkably still, thanks to an extremely supple neck, while the rest of the body is in motion, in order to allow it to focus on what it sees in exceptional detail.

In birds of prey, the eyes weigh more than the brain. The two eyes are twice as large as the brain itself. They need to be huge. They are packed with receptors, some types of which humans don't have at all. Like all birds, raptors have not merely two (as we do) but three types of photoreceptors in the eye. Because of this, birds are thought to be able to experience colors that humans cannot even describe. Their retinas, unlike ours, contain few blood vessels. Instead, a thin, folded tissue called pectin, unique to birds, brings blood and nutrients to the eye without casting shadows or scattering light in the eye as blood vessels do.

Most birds, like most mammals, have a single area within the eye of perfect vision, called the fovea, where cone cells, which detect sharp contrast and detail, are most concentrated. A raptor's eye has two foveae. One is for lateral vision, the other for forward vision. A human eye has two hundred thousand cones to each square millimeter of fovea. Sparrows have twice that. Raptors have more than a million.

Raptors see in such fine detail that humans need microscopes to begin to imagine it. They also have a wider field of vision than we do, thanks to the second fovea, as well as better distance perception than other birds. Most birds' eyes lie at the sides of the head so that when they look at something, they use one eye at a time. With forward-facing eyes, raptors have binocular vision like ours, but better. Fields of view of the left and right eye overlap, allowing the brain to compare the slightly different images from each and instantly calculate distance.

And there is something else about a raptor's vision, something more difficult to describe. "These birds don't think the way we think," Nancy tells us. "They don't learn the same way we do." Because of our differing brain circuitry, birds capture at a glance what it might take a human many seconds to apprehend. For all birds, but especially these, seeing is not merely believing; seeing is knowing. Seeing is being. That is what I see in Jazz's monstrous, devouring eyes: the windows to a mind completely different from my own.

"It's instinctive," says Nancy. "It's not spiritual. A falcon is at once the stupidest thing you'll ever deal with—and the most instinctively developed thing you'll ever deal with."

Instinct gets short shrift among most humans. We value thinking instead and dismiss instinct as the machinery of an automaton. But instinct is what lets us love life's juicy essence: instinct is why we enjoy food and drink and sex.

Thinking can get in the way of living. Too often we see through our brains, not through our eyes. This is such a common human failing that we joke about the absentminded professor or the artist so focused on his imagined canvas that he walks into a tree.

But Jazz won't smack into a tree. We are out in the field across the street now, and Nancy unclips the jesses that keep Jazz tethered to my glove. "Let her fly," says Nancy. I give Jazz a brief toss from my glove, and she sails into a pine. She looks down at us. Now I am worthy of Jazz's interest. She knows something is about to happen. For the first time, I am bathed in her sight. It's a baptism, and feels momentous, transforming.

"Now call her in," says Nancy. She takes a piece of cut-up partridge out of a baggie in her pocket and places it between the thumb and forefinger of my glove. "Jazz!" she calls. I extend my left arm and look up. A huge, powerful bird flies toward me.

Not everyone would like this, I realize. An exceptionally brave biologist with whom I have worked in Southeast Asia, hiking in search of bears among forests littered with unexploded ordnance, confesses he would be scared. It's a genetically programmed human reaction. Birds like this once hunted and killed our ancestors. A famous fossil hominid, the so-called Taung Child discovered in South Africa in 1924 and described by Raymond Dart, bears the marks of this predation. When I was in college, we had been taught that this long-dead australopithecine child must have been killed by an ancient leopard. Now, from careful reexamination of the skull, we know that the death blow dealt to the brain came from the talons of an ancient relative of the crowned hawk eagle—a raptor that still hunts large monkeys in the same manner today. Our kind has rightly viewed birds like Jazz with caution for more than 2 million years. No wonder so many people flinch.

But as Jazz's talons reach for my glove, my heart sings.

She lands surprisingly heavily for a bird. The squeeze of her talons is strong. Her piercing eyes are focused on the job at hand: tearing flesh with her beak and feet with an intensity that encompasses at once rage and joy. Though Jazz has done nothing more than land on my hand, I feel she has given me a great gift.

Sy with the Harris's hawk Jazz

On my hand, I hold a waterfall, an eclipse, a lightning storm. No, more than that. Jazz is wildness itself, vividly, almost blindingly alive in a way we humans may never experience.

This is one reason I have always been drawn to animals: their sharpened senses give them a fuller experience of the world. Largely oblivious to the symphony of scents, humans experience only a small part of life. We hear but a sliver of the range of the world's voices and have evolved to depend on vision most of all. But although we live through our eyes, birds do so to an even greater degree.

Birds' eyes gather more of life than ours do. Perhaps this is why I could feel Jazz so purely, densely full of life, filling up the moment—here, now, and nothing else. The Buddhists say there really *is* nothing else, because now is timeless; now is everything. Perhaps because of this, Jazz seems more immediately alive than any human I have ever known. To be in the gripping gaze of that bird is like looking directly into the sun. The class is a transforming experience. I am hungry for more.

* * *

Two years later: I never forgot Jazz. I longed for the laser intensity of her eyes, the assured passion of her instinct. I longed to be with her and know more about her kind. But life intervened. A pressing book deadline; an expedition to New Guinea; a new hatch of baby chicks; a rescued border collie; two national book tours; an expedition to Mongolia.

But I knew I'd return. And now I am back in Nancy's New Hampshire farmhouse kitchen, with its green checkered curtains and worn wooden countertops on which endless piecrusts have been rolled. With the cheerful air of someone baking cookies for her grandchildren, Nancy is using a bread knife to cut off the heads and legs of frozen baby chicks and popping out the hearts to use in training the new Harris's hawks.

"They love the heart," Nancy is telling me. "They also love the heads."

I try not to think the meat she's cutting was once sweet baby chicks, like the fluffy babies who sleep in my sweater and perch peeping on my shoulders and head. I am trying to stay focused on why I came here.

I came back for Jazz. Or so I thought. When I phoned Nancy to set up an appointment, I learned to my shock and sorrow that Jazz was dead. The majestic hawk had never shown a sign of weakness; her appetite was sharp, her flight strong, her feathers perfect. But one day, a year before I called, when Nancy went to visit Jazz in her aviary, the bird looked sick, and when she picked her up, Jazz died in her hands. A necropsy showed it was cancer.

I was so stunned and sad I couldn't speak. I was loath to let on to Nancy that Jazz had meant so much to me after our single half-day encounter. I knew I could not have meant anything to Jazz; I doubted that Jazz would remember me and wondered whether even Nancy would. But she did.

"Before you came two years ago," Nancy said gently, "I wasn't sure Jazz would fly to anyone but me." Previous attempts had failed. She just wouldn't land on anyone's glove but Nancy's—until mine.

Jazz had in fact given me a great gift. I would go forward with my study of this discipline, I decided right then, in her honor.

"Some people take to falconry easily, and others will never be a candidate for this," Nancy continued. "You need to read the bird. For some that comes hard. For some it comes easy. And frankly, Sy, it comes easily to you. You are working in the same time frame, flowing with what the bird is doing."

This is what keeps me from retching as Nancy cuts up the chicks and pops out their hearts. To flow with a hawk, to enter its timelessness: I want this so bad I can taste it.

Nancy and I have just come in from greeting the new crew. On this fine, cool October morning, I will be working with young Harris's hawks who hatched at a breeder's this summer. They're all just about as big as adults, and impressive birds. But one look in their eyes shows they're still just babies: their eyes aren't mahogany like Jazz's, but still a grey blue.

In the side yard are their aviaries, called mews. Each bird has its own separate red wooden building about twice the size of a spacious toolshed. Next door to Sabretache, the saker falcon, lives Smoke, the larger of the two Harris's hawk sisters. Long before Nancy even thought of teaching Smoke to chase a flying lure to prepare her for hunting, Smoke was chasing after leaves that flew off the trees. "She's going to be a wonderful hunter," says Nancy. Though Smoke is seven ounces heavier than her sister, her plumage looks less like that of an adult. Unlike Jazz's rich brown, Smoke's breast feathers are the color of mocha ice cream drizzled with melted chocolate.

Smoke's smaller, darker sister, Fire, lives in the mews across from her—next door to Moseby, a female goshawk, and two doors down from Banshee (who, Nancy tells me, has overcome her crabby adolescent stage and become a fine companion hunter). Fire is screaming bloody murder. Nancy calls the racket "peeping." Baby hawks do this when they want attention from the parents. "Fire's smart," Nancy tells me, "but she is stunted in her emotional development." Her sister used to steal her food when they were housed together at the breeder's, and this made Fire babyish and demanding.

Out back are more birds. Nancy only introduces me to two Harris's hawk brothers who hatched from different clutches but look identical. Scout hatched in May and is eight weeks older than Sidekick. But Sidekick is the fearless one. That's sometimes a problem. "This guy lands on my head, on my shoulder—and he talons me! He opens me up. Jazz used to grab me like that," Nancy remembers. "But he's a smart cookie. He'll learn."

Harris's hawks are smarter than other hawks, Nancy insists. "A goshawk will watch a rabbit go down a hole and just sit there," Nancy said. "A Harris's hawk will realize your dogs will chase the rabbit out of the hole, and he'll go wait for the rabbit to emerge on the other side." Harris's hawks are naturally companionable. Sometimes a female (as with all raptors, the larger sex) will take two mates at once. Like crows' young, those from a first clutch of Harris's hawks often stick around the parents to help them feed the second clutch. And unlike most other birds of prey, in the wild, Harris's hawks often hunt in packs. Cooperative groups of up to a dozen may gather to hunt hares, birds, and lizards in the Southwest desert, especially in winter. Harris's hawks are born genetically programmed for partnership, about which I am just beginning to learn.

"Doing is the best way to learn," Nancy tells me. Traditionally, she says, an apprentice learned by mostly just watching a more knowledgeable falconer. But today I'll be able to do more than just watch; with four young birds to train, she's giving me the great compliment of allowing me to help. "You'll fly a bird today," she tells me.

We enter Fire's mews. Nancy pushes her glove up to Fire's breast, and the bird steps onto her glove. Inches from Fire's ebony talons, Nancy places a clip through a hole in the jesses to tether the bird to a rope tied to the glove. But as soon as we leave the mews, Fire tries to fly. She launches—but because she is hooked to the tether, she flips over. She hangs upside down from the leather straps on her great yellow feet, thrashing the air with her huge, powerful wings. Nancy reaches toward her with her naked right hand and gently pushes on Fire's back to swing her back up to the glove—impressively without tangling the jesses, being bitten, or getting taloned. But Fire flies and flips again. Nancy again scoops her back to the glove.

This strange behavior is called "bating." Why would a hawk want to hang upside down from your glove by its feet? "They don't want to hang upside down," Nancy says. "They want to fly." Why don't they figure out that if they try to fly while they're tethered, they're going to hang upside down? "Well, they just don't." These are such strange creatures. How will I ever begin to understand what is going on in their heads?

Nancy has to weigh Fire to make sure she's not too full from her last meal to want to follow me. She urges the bird to step onto a scale. Fire weighs twenty-six ounces and is hungry. Nancy unclips the line tying her to the glove and lets her fly onto the branch of a pine in the side yard. Nancy tells me to walk away with my back to Fire, and then turn around to see if Fire is watching me.

She is. I can feel Fire's gaze even before I turn.

I take perhaps twenty steps before Nancy tells me to call the bird. Surreptitiously I reach my ungloved right hand into the pocket of my jacket to part the lips of the baggie hidden there. My fingers find the cold, bloody slime of a cut-up chick. I feel for the point of a beak; I want to offer the coveted head. I place it between the thumb and forefinger of my borrowed glove. I turn toward Fire and extend my arm.

I desperately want to do everything right and remember all that Nancy's told me.

"Present the back of your shoulder to her," Nancy had said. I turn so that my back is to the hawk, extend my gloved hand, and look at the hawk over my left shoulder. This is a stance that could protect me in case I ever fly a goshawk. Goshawks not infrequently attack your eyes, which is how many hawks naturally kill their prey, forcing their talons through the skull via the eye socket into the brain, Nancy explained. That's why "if they think there is any chance the food might go away, they fly into your face and go for your eyes."

Nancy learned this firsthand with a goshawk who belonged to her husband, Jim, from whom she learned falconry when she was his apprentice. It was their wedding anniversary, and Nancy was all dressed up waiting for Jim to get home from Boston to go out to dinner. Arriving late, he asked if she would feed his new goshawk while he showered and dressed. It would be easy, he told her: "Just present the first chick on the glove. Let her eat it. Present the second chick. Let her eat. Present the third. Then you're done."

"So," Nancy told me, "I go into the mews, she's sitting there. I present the chick. She eats it. No problem. I do the second one. Fine. I give her the third and I turned sideways to leave—and I felt talons going around my eyeball!" She had two puncture wounds on her face, both bleeding copiously. "If she had clasped her talons," Nancy says, "she would have torn my eyeball out."

Jim found his wife leaning up against the door of the mews, bleeding and furious—not at the bird, but at her husband.

"What didn't you tell me?" Nancy demanded.

"I told you to present three chicks . . ."

"*What didn't you tell me?!*"

And then Jim remembered. "Oh, I forgot—after the third chick, I reach out and go *fluffel, fluffel, fluffel* on her breast feathers."

Just that one change in routine was enough to provoke an attack. "That's how much of a raw nerve they are," Nancy said. If things aren't going as expected, the best plan is to attack. "They're acting on an instinct that is so fast, it's like lightning in a bottle."

That's why she trains her students to present the back of the shoulder to any incoming hawk, she told me. "Should you ever fly a gos, and it decides to go for your eyes, you can turn and shield your face with your ungloved hand."

This is what you are dealing with when you work with a bird of prey. Though most species are far less aggressive than the goshawk, all of them can be unpredictable and dangerous. "It's like handling a loaded gun," my husband, Howard, said with alarm.

That lesson is not difficult to remember. But I am not thinking of this as Fire flies to me, talons reaching for my glove. I am so hungry to have her land on my fist, there is no room in my heart for anything else. My whole soul feels like a yawning hole that only this bird can fill. She lands on my glove—*smack!*—and begins tearing at the chick head. I pull my arm in to draw her closer to me.

Were I to find a crispy drumstick on my dinner plate, I would feel ill. But as she eats the meat, her joy is mine.

What is happening to me?

Weeks progress. Thanks to Nancy's clarity and patience, slowly I am beginning to know how to move, how to watch, how to think around hawks.

At first, it seems like a lot to keep in mind. As in all new endeavors, there are strange little injunctions to remember—like never say *Shhh!* to a hawk. (It's similar to the warning hiss they give before biting.) There's a lot of equipment to learn how to use—never my strong suit. There is a certain amount of technique—how to hold the bird (always higher than the rest of the arm), how to "load" the glove (without letting the hawk see you remove the bait from your pocket), how to attach the tether to the jesses (quickly, before the bird "foots" you—the falconer's term for when a bird tears its talons through your flesh).

Getting things right really matters. If you make a mistake, you can be hurt. But far worse, your bird can die. A hawk's wings, tail, feet, and eyes are so delicate they can be easily injured when you take the bird out of the mews. Even an unexpected wind can lift, twist, and sprain a wing when a bird is tethered to your glove. You must be careful, especially with Harris's hawks, who are native to warmer climes, not to fly a bird on a cold day. Even mild frostbite on the feet can kill. Infections can be lethal. Tangled jesses can break a leg. The list of hazards goes on and on.

Because they have regular food and access to veterinary care, hawks used in falconry generally live much longer than those who are free: 80 percent of wild hawks die in their first year; a falconer's might live twenty years. But even with human care, a hawk's life is fraught with danger. One of the greatest heartaches a falconer can know could happen each time you take your bird out of its mews: a wild hawk in the sky might attack and kill your bird.

Nancy knows this sorrow.

She was out hunting with a young peregrine falcon named Witch. Nancy had raised her from a downy, helpless chick. The name was short for Kitchen Witch, because during her chickhood, during the day, the kitchen was where Witch stayed, to be near Nancy all the time. At night Witch slept by Nancy and Jim's bed. Young raptors are irresistibly ugly babies—fat-bottomed tripods of fluff with unnaturally huge feet who chase Coke cans across the floor, scream at you with eyes full of wonder, and then fall on the floor, face-down, legs out, asleep. To raise a bird from this endearing, silly-looking chick to a powerful hunter demands a huge investment of love and work. And now Nancy and Witch were out hunting, cementing their partnership.

Witch flew over an orchard after game—and flew back with a red-tailed hawk in pursuit. Again and again the big red-tail hit the young peregrine. Witch landed in a tree, and though Nancy tried to call her to the glove, where she would be safe, Witch was too young and afraid to chance flying low—the higher hawk always has the advantage. Nancy raced off to try to get her car, parked far away, to get closer to Witch. She stood in the street and flagged down a logging truck to hitch a ride. Within twenty minutes she was close to the last tree in which Witch had landed. There, on a low branch, sat the red-tail. It looked much fatter than before.

For five days, Nancy returned to the spot, calling for Witch, swinging a lure baited with meat till her hands were bloody. But she knew what had happened. The red-tail had eaten her bird.

Once, as she was cutting up baby quail for the hawks, Nancy told me her motto, one she credits to a fellow hunter, Teddy Moritz: "Hunt hard. Kill swiftly. Waste nothing. Offer no apologies."

"It's nature's way," she says. And when Witch was killed, that, too, was nature's way. But Nancy still mourns.

Slowly, I am starting—just starting—to understand who birds of prey are.

Sometimes they seem shockingly mindless. By my fourth lesson since meeting Jazz, it's clear Fire knows me. She doesn't move away from me when I approach her perch. She eagerly steps onto my glove. Yet as I carry her toward the door to go outside, almost invariably she launches off the glove and hangs upside down from her feet, "bating." She can't possibly enjoy this—yet she does it again and again. Why can't she seem to remember the consequences and just wait till I unhook her?

Because, Nancy explains, so much of hawks' behavior is hardwired, instinctive. Fire bates off the glove because the urge to fly is overwhelming.

Equally overwhelming is the reaction to clench. A hawk who has caught a rabbit may not let go until the rabbit dies—or until the hawk does. If a rabbit races into its hole with a hawk on its back, the hawk will allow itself to be taken underground—and rather than let go, it will die there. Why? "It can't let go," Nancy explained to me. "If something wiggles, they bind their feet to it as a reflex. They don't have the choice. The mind is not involved in this at all."

If you are footed by a hawk, you must remember this. One of Nancy's apprentices, Rita Tulloh, discovered this when her huge red-tail hen, Scarlett, nailed her hand. Your instinct is of course to pull away, with the bird's talons deep in your flesh, but you mustn't struggle, or else you will never extract yourself from its grip. "The clench doesn't even have to go up through the thinking part of the bird's brain," Rita told me. "It's like a sneeze or a blink. Meanwhile, you just wait and drip." Rita will always have a scar.

But in other ways, a hawk's mind is like a steel trap.

"They have the instinct to hunt and fly," Nancy told me, "but *how* to hunt, they learn. It's as if there's some file folder in their heads about hunting success that they learn and never forget."

When it comes to hunting success, hawks learn instantly—and once they do, they remember it forever. Nancy told me about the first time Jazz went hunting. She followed a pheasant sixty feet up in the air, grabbed it, caught it—and then had no idea what to do. She let go and was frustrated and mystified when it disappeared.

"But she never let that happen again," said Nancy. The next time she grabbed a pheasant, Jazz held on and killed it. When hunting, a hawk doesn't make the same mistake twice.

They are not automatons, however, and can be quite emotional. Nancy told me about the winter day she was hunting with her first Harris's hawk, Indy. She slipped on the ice from a small bridge and fell into a shallow stream. Indy was unhurt, but he was furious: "He thought I had thrown him down!" she said. He screamed insults into her ear and remained angry with her for a week.

Anger and frustration tend to be the emotions they most often show. (Oh, great—a loaded gun that's mad at you.) At least this is one emotion that's easy to notice: I had no trouble recognizing Scout's disapproval when he saw me wearing a black headband for the first time. "WRAAACK! WRAAACCKKK! WRAAACCKKKK!" "WHO THE HELL ARE YOU? GO AWAY! GO AWAY!" He screamed at me like he wanted to kill me. He probably did want to.

I am learning how to read these birds' postures and gestures, understand when they feel comfortable and when they do not. I know when Fire is going to bate off my glove and how to deal with it. I remember that Scout doesn't like it when I present my shoulder, so I present my chest (remembering not to do this if I ever fly a goshawk). I know that Smoke and Fire both recognize me by now and are not annoyed by my presence. I feel confident handling their jesses and even touching their feet. If I can learn how to behave correctly, perhaps these gorgeous creatures—whom I love like an Aztec loves the sun—will learn to trust me not to do something scary or stupid.

One day, before an event at which Nancy would fly Fire before a crowd, I held Fire while Nancy gave her dirty tail a bath in the kitchen sink. I held her jesses firmly between the thumb and forefinger of my glove, expecting she would bate or even try to bite or foot me. To my surprise, she was calm, and I felt a wave of tenderness sweep over me for this powerful predator during a lovely, intimate, and quiet moment. I think she rather enjoyed her bath.

But I know not to look for affection from these birds—now or ever. Falconers, each of them crazy in love with their birds, accept that their birds will never love them back. At least not in the way a fellow mammal shows you its love. From these birds, I want something else—if I can earn it.

"They may show you a certain companionship. They can become comfortable with you. With Indy," said Nancy, "we knew each other so thoroughly, I could close my eyes and know exactly where he'd place himself in a tree relative to me. It's a beautiful partnership. But if you break their laws, you'll pay."

"If you want love out of this," Nancy tells her students, "you're too needy. Don't be a falconer."

For a human to love without expecting love in return is hugely liberating. To leave the self out of love is like escaping the grip of gravity. It is to grow wings. It opens up the sky.

Nancy is teaching me how to work with the lure—a tool to help young hawks learn to chase and catch prey on the wing. The lures with which we work are leather, one in the shape of a rabbit, the other shaped like a bird. They're baited with meat. Thrown to the ground or swung on a rope in the air, the lure inspires pursuit and helps the young hawk practice maneuvers to attack and seize its prey.

Nancy takes out Scout to show me how it's done. She swings the flying lure, baited with a chick torso. He makes a pass but misses, lands in a pine tree, and watches. She swings it again. He misses twice more. He screams. He seems to be tiring, nearing frustration. He wants to catch it but hasn't the skill. The fourth time, Nancy swings the lure as slowly as possible—and he seizes it in the air and tackles it to the ground. He spreads his wings and tail over the prize, mantling, as he tears at the meat. It is an act so intimate, private, and selfish it feels as if I ought to avert my eyes. But of course I cannot take my eyes off the bird. Nor do I really want to.

Nancy shows me how to retrieve the lure. While the bird is eating, you step behind the bird, onto the lure, taking care not to hurt the tail feathers; you load your glove with bait, but keep it hidden; and when the bird is finished eating, show it the meat on the glove. The second the bird flies from the lure to your glove—again, while it is occupied with eating—turn your body to retrieve and hide the lure. Nancy stuffs it in a back pocket of her vest and attaches the bird's jesses to the leash.

It looks simple. Now I'll try it with Fire.

The bunny-shaped lure is baited with a chick wing. I toss it onto the driveway. Fire flies to it and begins mantling as she eats it. I am lost in her pleasure, lost in her beauty, drowning in my love of this bird.

I forget to load my glove. Nancy reminds me. And then I forget to step on the lure. Quickly Fire finishes eating, and unwisely I show her my baited glove. Up she flies to my fist—her talons still firmly grasping the lure. "Oh, that's bad," says Nancy. Indeed it is—now it will be difficult to detach her feet from the lure. Nancy has me put the bird, glove, and lure back on the ground—now I am thoroughly flustered—and again shows me how to do it right. Fire now flies to my glove and leaves the lure behind.

Later we discuss why letting Fire bring the lure to the glove was such a big mistake. Between the string to the lure and the string to the glove, Fire could get dangerously tangled. Trying to escape from the tangle would only cause her to tighten her grip on it the more. "Plus," I said, thinking I could apply a lesson from dog obedience, "allowing her to take the lure to the glove and getting two bits of chicken rewards her for a behavior you don't want to reinforce."

Nancy corrected me emphatically. "You must never think of rewards or punishments," she told me. "If you think in terms of rewards and punishments, you're not thinking partnership. They don't serve us. We serve them."

In falconry, you don't train a hawk to do things for you. In *A Rage for Falcons*, Steve Bodio puts it this way: You train a hawk "to accept you as her servant."

"We are not giving them anything," Nancy stressed. "It's either theirs or it's not. We are working with the bird. It's a partnership. And even after all these years, I'm the junior partner. I am not the master."

This is the biggest mistake I have made so far, and I am feeling exceptionally clumsy. My hands won't work in the cold. My glasses, which I wear for distance, blur close-up work, like linking the French clip to the hole in the birds' jesses. But as the day's lesson is about to end, Nancy surprises me with a new assignment.

"I want you to invest in a whistle. To call the bird. The whistle means food. It will make you more independent." That will mean stopping by the hunting goods store—how odd, I think: me in a hunting goods store.

"And I want you to order your own falconry glove." It must be leather— nothing else will protect against those talons. I never buy leather, but eagerly I find myself taking down the number for the falconry supply outfitter.

"You'll have the whistle," Nancy continued. "You'll have the glove. There's only one thing missing."

"What's that?"

"The bird."

Nancy is inviting me to be her apprentice. This is a huge honor, and an act of great generosity on her part. It means Nancy can no longer collect the checks I write out for lessons. If I am her apprentice, she teaches me for free.

But the apprentice makes the larger investment. It's a two-year commitment—and usually leads to a lifelong passion.

First you must complete a weekend hunter's safety course. Pass the exam for a hunting license. Build a mews and set it up with all the furniture a hawk would need. And then, sometime between September 1 and November 30, catch a young wild red-tail and make it your own.

Could I do this?

There are several ways to catch a wild hawk. One way is to prepare a trap called a bal-chatri. This ancient Indian device looks like a loosely woven upside-down basket, covered with small nooses. You can make a bal-chatri from slats of cane, hardware cloth, or kitchen wire—for a small bird like a kestrel, even a kitchen strainer will do. Beneath the trap you place a live rodent or pigeon to attract your quarry. Place it in an area you have seen young birds frequent, and go off and hide in the bushes or in your car. When the hawk lands on the bal-chatri and tries (unsuccessfully) to grab the bait, its feet tangle in the nooses. You then rush out from your hiding place, throw a towel over the hawk, disentangle the feet, and take your captive home.

Capturing the bird isn't always easy, but that's not what worries me. My concern is this: could I in good conscience take a bird out of the wild?

Other committed conservationists do. After all, a falconer's bird will almost certainly live longer than a wild one, and once your apprenticeship is over, you can set it free. Having made it through that first, most perilous year, the hawk will have an even better chance of survival than if it never met you.

But what about the Ladies? And not only our beloved hens, but now Elizabeth's Rangers! How could I keep a hawk on the property with chickens, too?

Even if I managed to work with the hawk away from the chickens, even if I set her free after a year, wouldn't she find me? I imagine looking up into the sky to see my hunting partner coming back to me, wild and free—and then slamming to earth like a lawn dart onto the body of Pickles or Matilda or Peanut. One glimpse of the chickens, and they would forever be filed under "hunting" in the hawk's steel-trap mind—causing her to return again and again, until every one of our beloved, gentle, harmless hens had been killed, pierced through the brain by her cruel, perfect talons.

I ask Nancy.

"The hawk would kill the chickens," she tells me flatly.

I pose the question that night to friends over pizza at a restaurant. Could I be a falconer? One of them is the author-illustrator Lita Judge. Her grandmother, Fran Hammerstrom, a legend among falconers and birders alike, was a pioneer in banding hawks and rehabilitating injured birds, and the first to successfully breed endangered golden eagles in captivity. Lita grew up with hawks and owls swooping through the rooms of her grandmother's house and knew how to handle them all. What does she think?

Lita hasn't gone near hawks or owls for years, she says. She has purposely avoided working with them. "I'm afraid it would take over my life." Falconry, she warns, is addictive.

"One thing you might not want in your life," she says, "is lots of dead things in your house all the time." My husband puts down his pizza and rolls his eyes.

Lita looks at my hands. "Also, the skin on your hands is pretty thin. There's not much flesh there. A hawk could give you a pretty disabling injury. If you got footed, it would go right to the bones and the ligaments and tendons." Some eagles used in falconry, the golden eagles of Mongolia, can crush and break the bones in a human hand with a slight flex of their talons. "That's a consideration for you," Lita says, "since you take notes with your right hand."

But what about the chickens?

Lita doesn't even have to think about that one. "The hawk would kill the chickens, all right. No question about that."

Could I find some way around this? Could I keep the hawk somewhere else, off our property? My friend Liz Thomas, who lives seven miles away, generously offered her home, which is inhabited by two Australian cattle dogs, two parrots, several cats, and one husband—but no hens.

But then there is the issue of my travels. Besides short trips for speaking engagements and book tours, I am often away for a month or more at a time working in some remote jungle or desert. In my absence, my husband and neighbors have, over the years, cared for a menagerie of critters ranging from our 750-pound pig to hens, ferrets, dogs, and parrots. But in good conscience, can I ask another person to tend a loaded gun?

As I tell friends about learning falconry, many are excited for me, but as I describe it further, some grow visibly distressed. One, a biologist, now eighty, who flew a sparrowhawk at thirteen and likes to tease me about my vegetarianism, asks with concern: "*You?* A hunter?"

Our tenant Elizabeth, the Chicken Whisperer, is too kind to say anything judgmental, but I can see it on her face. She hates hunters.

Many people, knowing how I love animals and eschew meat, assume that I hate hunters, too. But my mother was a hunter. Growing up in rural Arkansas, she hunted squirrels and ate them. My friend Liz, who had offered her home to me should I get a hawk, is also a licensed hunter—though she hasn't actually shot a deer. Other friends are hunters, too, many of them tireless conservationists, people I deeply admire. Hunting is a far more humane and ecologically sane way of obtaining meat than factory farming, which is hideously cruel, wasteful, and, because it generates tons of concentrated animal waste, ecologically disastrous. I understand hunting for food. But unlike a cheetah or a polar bear or a falcon, I don't need meat to survive. So, rather than taking up hunting, I gave up meat.

But falconry, I realize, is hunting. Venery. For years, as a child, I had believed that venery was one of the seven deadly sins, like envy, greed, and sloth. But no, *Webster's* assures me I was wrong. Venery is defined as "the art, act, or practice of hunting."

A second meaning, however, is listed in *The American Heritage Dictionary.* "Venery: the practice or pursuit of sexual pleasure." The word is rooted in the Roman goddess Venus, whose name meant desire or love.

"Gaia put the will to hunt deep in our psyches," my friend Liz, whose family was the first to study Bushmen in the Kalahari, observes in one of her many books on people and animals, *The Hidden Life of Deer.* We have hunted throughout human history; even our closest relatives, the chimpanzees, go hunting. By their excited hoots and thrashing displays, they show how thrilled they are when they are successful. Hunting is at the heart of life for many animals. I wanted and needed to understand it.

Could I be a falconer? I weigh my options. On one hand, I have my hens, my home, my marriage, my job. On the other hand, my desire.

On a freakishly warm, cloudy morning in October, our third class after meeting Jazz, Nancy tells me about yarak.

It's a word that is difficult to define, Nancy says. Its origins are obscure. *The Oxford English Dictionary* says it might come from the Persian *yaraki,* meaning power or strength, or from the Turkic, for the proper heat for tempering metal. Falconers speak of a bird being "in yarak"—in proper condition for hunting. But yarak is not about physical health or strength. Yarak is instead something central to the psyche of a bird of prey.

The bird who taught her about yarak was her first Harris's hawk, Indy, a gift from a good friend who breeds hawks and rehabilitates eagles. A big, strong bird, Indy was a powerful hunter from the start. This was back when Nancy had just finished her falconer's apprenticeship, before she and Jim had moved to rural Deering, where they live now. They were then living near a large housing development, where wild game was scarce. She offered Indy captive-bred live quail and partridges; in the fall, Indy also hunted migrating birds.

The winter started out mild. But then came a storm. The migrants left. And then Nancy ran out of captive quail and partridges.

But Nancy wasn't worried. She flew Indy frequently from tree to glove. He was eating plenty and getting exercise and company; he just wasn't hunting.

Then one day she was with him in his mews, changing his jesses. He reached out with his foot and circled her wrist with his toes—and held on. "He didn't hurt me," she said. "But he was dominating me. This is called 'braceleting.' But I didn't know it at the time." She finished with the jesses.

Two days later she made a mistake. She was changing the jesses again and unwisely changed the one farthest from her first, leaving the leg closer to her free.

Indy exploded. "It was as sudden as a windstorm," Nancy said. Without warning, the bird screamed at her and flew at her face with his talons.

He hit so hard she felt she had been punched in the mouth. He sliced her so deeply that fifteen years later, you can still see the scar: a thin white line running from her left nostril to her upper lip. "By the time I finished changing the jesses," she said, "the blood was flowing so hard it was covering my hands."

This wasn't crabbiness, like Banshee. This was an attack. "It was like he'd never seen me before," she said. "I think what he did surprised even him."

The attack made no sense to her. Harris's hawks are supposed to be cooperative—"but this was not cooperative," she pointed out. What was going on?

Her husband, Jim, with whom she had apprenticed, said it might not happen again. She asked the president of the North American Falconers Association. "I don't know," he confessed, "but Indy might be as upset as you are." Then she called Frank Beebe, a falconer of international repute, author of many books on hawking. "I have a feeling you need to think back," he told her. It sounded to him like a buildup of yarak—the often explosive buildup of hunting desire.

Nancy went over in her mind the events of the previous six days with Indy. It had all started, she realized, when she'd introduced him to game—and then cut him off.

"I had not realized it was like flipping a switch," she told me. "When you cut it off, it's like the power goes off and the oil burner floods. You have to turn the switch off before lighting it again, or BOOM! I hadn't turned the switch off. The hunting drive wasn't going off."

Yarak names this drive, this desire. It is the bird of prey's greatest earthly pleasure and its deepest frustration, twined tight. Frightening and beautiful, yarak is rapture and longing, love and death married in one timeless moment. If the eye is the mind of the raptor, then yarak is its wild soul. Yarak is wildness incarnate—dangerous and delicious and pure. Yarak names why I long for communion with birds of prey. This is as wild as wildness gets. There is but one way a human can touch a hawk's wildness: to join it, as its partner, on the hunt.

One day Nancy phones me at home. Usually we set up the next lesson at the previous one, or we communicate by e-mail. The phone call feels momentous.

"I want you to see the end result of what we do," Nancy tells me. "Up to now, you've been taking baby steps. It's great for you, but very unsatisfying for the birds.

"The whole thing about falconry," she says, "is getting to hunt. Not to hunt with a falconry bird is to deny what they are. Hunting is the strongest instinct they have. It overrides migration. It overrides procreation. Hunting game is what cements the partnership. And you will be part of that successful hunting file."

She is inviting me to witness a hunt. Tomorrow.

"I want you to see in the field how these birds respond to everything I'm teaching you at the house," she says. "It's magic."

My heart pounds. I am hungry for this.

The day's hunt will involve, Nancy said, something called a quail launcher.

Immediately a shiver of horror goes through me.

Nancy asks gently, "Would that upset you?"

"No." I desperately want to go.

The next morning Nancy picks me up in her truck. In the back, from their travel crates, the hawks Smoke, Sidekick, and Scout are shrieking like sirens, and from within a portable kennel, a female German wirehaired pointer, Stormy, whimpers with excitement. We are heading to a hunting club, which stocks its extensive fields with captive-bred pheasants and quail. Nancy outlines the plan for the day.

"What I'm trying to do," she says, "is first take Smoke, who already works with dogs. With Stormy and Smoke, we'll do one sweep of the field to see what the dog turns up. And then we'll do the quail launcher."

Two quail to be "launched" will be awaiting us in a box, courtesy of the hunting club's owner. The launcher is not, to my relief, some sort of cannon or catapult that throws quail. It's more like a folded-up trampoline. The act of being launched does not hurt the quail, although it is surely frightening. Almost immediately the quail begins to fly on its own.

So why the quail launcher? For the falconer, the contraption is a teaching tool to help a young hawk learn that the dog is part of the hunting team—which left to its own devices, the hawk would not discover. Harris's hawks naturally hate dogs. In the wild, they hate coyotes so much they scream at them. This is exactly what Smoke will do when she catches sight of Stormy—in the process, alerting any nearby prey that a hawk is near and ruining the chance of a successful hunt. The hawk will never get the chance to see the dog put up game.

Using the quail launcher can help the falconer control the situation. Done correctly, the quail can be launched while the bird is looking in the right direction to see the dog on point. "Once the hawk realizes the dog puts up game, he ceases resenting the dog," Nancy said. "He'll start to take cues from the dog. These birds, their instinct to hunt is so strong, they'll never forget what constitutes hunting success—and now the dog will be included."

We are driving to the field where we'll ask Stormy to look for pheasant with Smoke. I'm scanning the sky and the treetops, as I always do—looking for hawks.

"Look who's here," I say to Nancy. There's a male red-tail in a tall, leafless maple.

Nancy knows him. "I've had him hunt my birds," she says. She gets out and lets the dog out, too. "GO AWAY!" she shouts at the wild red-tail. As if obeying, he flies toward a pond, floating on the strong wind, circling—and then dives down, dropping like a stone. "He went down with a purpose," Nancy says. He saw a pheasant. There's game out here.

But finding it won't be easy. It's windy. The wind interrupts the scent the dog needs to follow. "This is a hard day," Nancy says.

The plan is to walk back to the cover at the end of the road, where pheasant might be hiding. Smoke steps out of her carrier. From Nancy's glove, she flies to a tree. She can better watch the dog from there.

"Stormy—hunt 'em up! Find bird! Find!" Nancy says to her brown-and-white dog. Stormy is glad to oblige. Nancy and Stormy have been hunting together for nearly nine years, ever since Stormy was "the puppy from hell."

"If it was something that would be featured on the *Antiques Roadshow*, she'd chew it," said Nancy. "But that's what makes a great hunting dog—a smart, active, curious puppy."

Stormy's sniffing everywhere, zigzagging back and forth, searching for the edges of a cone of scent that will lead to the game. The large brass bell at her collar—both dog and bird wear bells to help us keep track of them—chimes with excitement.

"Stormy's getting birdy," Nancy says under her breath. And now the dog beelines. She's tracking now. And from her treetop perch, Smoke is watching, so intensely I can feel her gaze burning, hot as a laser. Smoke shakes her tail. It's a sign she is getting ready to fly—either to a new perch from which to better observe Stormy or after game that she has spotted herself.

We're walking through fields rough with the dry brown seed heads of goldenrod and ragweed, the thorny stems of wild raspberry. The white fluffy parachutes of milkweed are strewn everywhere like feathers. Beneath the thick cover, there are holes to twist your ankles, and at face level, branches to poke your eye. But most of all, if you are carrying a bird, you must guard the safety of its eyes—not yours.

Filled with tension, we're all watching, trying to see everything at once. We're watching the dog's posture. We're watching Smoke's attention in the tree. We're watching for pheasant. And we're watching out for the red-tail, should he reappear. We are not the only hunters here. Smoke could be killed, just like any wild hawk, just like any wild prey. This is a threat that all wild birds—hawk or dove, crow or pigeon—must live with at every moment.

Nancy wants to call Smoke to her glove, where she'll be safe. She takes a chicken leg from her pocket and holds it up to call her. But Smoke isn't hungry for food; she's hungry for chase. She is ruled by yarak. She keeps her eye on the dog.

Now the tinkle of Stormy's bell stops. Just above the ragweed, we can see her spotted back go rigid. "Look what we've got!" cries Nancy. "A point!" But a second later: "Oh, damn!" Sure enough, the red-tail's back. He soars above us; soon it's evident that this hawk, too, has seen what Stormy has pointed out.

The pheasant bursts from cover. "Get it! Get it!" Nancy cries urgently. Smoke flies after it. She misses and lands in a tree.

An eager hunting partner flies to master falconer Nancy Cowan's glove.

In her short lifetime, Smoke has not yet killed a pheasant, but she has chased three. Her first time was just three weeks ago, right here. Smoke was on Nancy's glove. Right in front of her, Stormy got a point, a hen pheasant flew up, Smoke chased it. And from that moment on, Smoke understood that the dog was an integral part of the hunt.

Now, the pheasant flies toward a distant pond, and the red-tail flies after it. Smoke wants to follow, too. But the red-tail is a real danger to her now. Smoke takes off.

"SMOKER!" Nancy yells at the hawk, furious, frightened. "Don't go there! I know the pheasant went there . . ." She blows the whistle. Smoke circles, lands in a tree, and looks down at us, disgusted. "She's mad she missed that pheasant," says Nancy. We'll try for another. "Stormy!" she commands. "Hunt 'em up!"

But the dog, too, is frustrated. She can't find another scent. She heads toward the tree where Smoke is perched.

"NO!!" Nancy yells. She knows what is happening: Sometimes if a dog can't find a pheasant or grouse, it'll go for the nearest bird—the hawk. And that makes the hawk furious: "Don't you point me, you moron!" Once Stormy pointed Jazz, and the hawk, enraged at the dog's stupidity and overcome by yarak, shrieked in fury and frustration and flew at Stormy with her talons. Nancy had to extract Jazz's feet from the dog's snout.

It seems inevitable: it will be a bloody, dangerous day. Yet I am not disgusted or distraught. I am living through my eyes now.

We've been searching for a good twenty minutes when, at 10:00 a.m., Nancy announces, "We've got a point!" Again Smoke is watching the dog intently. "She knows Stormy is birdy. Sometimes the bird sees 'em and wants the dog to flush 'em. Eventually they figure out a way to communicate that," Nancy says.

We're tromping over rough territory. Later I find that my pants are torn, my legs scratched and bloody. We're both sweating with exertion on this cool late autumn day to keep up with dog and bird. "You see why it's called 'hunting' and not 'getting,'" says Nancy. "You are seeing the essence of the hunt. Even if she doesn't get the game, the slip on game"—the chance to chase it—"is valuable."

And that is all Smoke will get today—a chase, but no catch. Nancy lets her attack the baited lure to satisfy her yarak and puts her back in her carrier. "Raptors are never happy," says Nancy. "They are always wanting, wanting to hunt. And if they are full, they are just flatlining." They are content, perhaps—but really, it is like their souls are in storage, awaiting the next chase.

At 10:50, we go to pick up the two quail that the owner of the hunting preserve has set out for us. He has left them in a little pet carrier in the garage. I look inside and my heart melts. There is nothing more innocent and appealing than a quail, with its rounded profile and soft brown plumage and black button eyes. One of the favorite books of my childhood was *That Quail, Robert,* about a quail who lived in a house with her people. Robert the Quail was so well loved that when the wife's diamond popped out of her setting and Robert swallowed it, the couple picked through quail droppings for weeks rather than sacrifice the bird. (Robert, in fact, kept the diamond, as grit to grind food in her crop.) I love the dear question marks on the heads of wild quail in Arizona, curled crests of black feathers. Watching quail chicks parade after their mother, in single file through the grass, I had been overcome with anxiety for their safety.

And now, I've come to watch these quail be killed. In their pet carrier, the two birds are still as stones, their stillness a fervent prayer that we somehow won't see them.

At least the quail launcher looks less terrible than I'd expected. It's an oblong box made of folded cloth, activated to spring open by remote control. Nancy positions the launcher carefully, then enfolds the quail inside like an astronaut in a space capsule. Stormy is ready, eager. Scout is still in his carrier in the car. But something goes wrong. The electronic launcher goes off prematurely and the quail launches to freedom. The little bird flies away and disappears into the brush before hawk or dog sees it.

I am flooded with relief. And then I feel a pang of guilt. Whose side am I on?

We have one quail left. My heart is pulled in two directions. Nancy places the little bird in the launcher. Stormy rushes over to smell it. The dog could trigger the launcher and release the quail before Scout is even on the glove. "BAD DOG!" Nancy yells. Stormy comes to her side, chastened, and I hold the dog by the collar until Nancy has everything in place.

"WRAAACCCK!" Scout cries on Nancy's glove. He's in yarak. He wants to hunt. He hates the dog. "WRRRRAAAAAAK!" he screams at Stormy: "You stupid dog! I HATE DOGS! I HATE YOU, HATE YOU, HATE YOU!" He shakes his tail, wanting to fly.

"We'll let Stormy go in and point," says Nancy, "and then Scout will notice something more important than hating the dog is going on."

But after Nancy had yelled at her, Stormy understandably doesn't want to go anywhere near the quail launcher now. The dog is everywhere but where Nancy wants her, obviously avoiding the device. "WRACKK! WRAAAAACK!!" calls Scout. He's extremely agitated. I can feel the buildup of frustrated desire. I worry he might bite or foot Nancy.

Now the dog is "getting birdy"—but not at the right bird. Stormy has found the pheasant—and wants to follow its scent instead of pointing at the imprisoned quail. "Storm-a-thon!" Nancy calls, urging her toward the launcher. "There! Go there!" Finally, the dog points. Scout is watching, and his attention is palpable, hot as a laser beam. Nancy presses the red button on the remote control. The quail shoots fifteen feet into the air and starts flying.

In a split second, Scout rockets off the glove, flips upside down, and attacks the quail from beneath, grabbing its belly with his talons. "Oh!" I cry. At that moment, there is no room in my soul for the quail's pain and fear. I am flooded with the hawk's elation. I feel it like a drug in my bloodstream, the ancient thrill of hunting success. I have never felt this before, yet it feels as familiar as my own skin. Through this bird, I have touched something very old and very wild, something I thought I could never feel: yarak. I have no desire to hurt that quail—but I realize that I want, more than anything, for this hawk to catch it. In the process, I have also learned something important about the nature of love. Now I understand why my father, an animal lover, bought my mother all those fur coats: his love for her blinded him to everything but her fire-bright happiness.

"You did it, Scout! Good bird!" Now he's got it; he knows how to hunt. Nancy and I are proud. But as Scout flies off with his prize, our elation turns to anxiety. We could lose him. Indy once caught a chuckar partridge here, disappeared with the prey, and wasn't seen for three days. We race after the bird.

Scout has flown toward the forest, on the other side of a hedgerow. If he has landed, we can't see where. And now overhead, against approaching storm clouds, another soaring shape draws Nancy's attention.

"Oh, no," says Nancy. "Is that a gos?"

"It doesn't look like a red-tail," I say.

Carrying the quail makes Scout's own body an easy and attractive target for a more experienced hawk. Even though a gos is smaller than a Harris's, "whoever is lower than you is easy prey," Nancy reminds me. It could easily kill Scout.

"Hold the dog!" Nancy says. If Scout has landed on the ground, the dog might scare him off. I hold the dog by the collar while Nancy races ahead. I'm grateful to remain behind. We know that Scout has caught the quail, but we don't know that he has killed it. In her pocket Nancy carries a sheathed knife for such an occasion: if the hawk is eating the prey but it is not dead, to spare its suffering, she will cut the head off the living quail.

I still visit Nancy, Fire and Smoke, Scout and Sidekick and take lessons from time to time. With my traveling schedule, I can't possibly take on a falconry apprenticeship at this point in my life. And then there's the not-inconsequential matter of my marriage and our Ladies. But Nancy and her hawks have profoundly changed me. Now, even as I eat my broccoli and my vegetable lasagna, even as I pray and work for compassion toward all sentient beings and write my checks to humane causes, I understand the falconer's mantra. People do not have to hunt, but hawks do; and these words name the unspoken rules by which hawks everywhere live their innocent, incandescent, wild lives:

"Hunt hard. Kill swiftly. Waste nothing. Offer no apologies."

Pigeons

Birds Find Their Way Home

I'm often uncomfortable at formal occasions, but I feel especially so at this wedding. I don't know a soul—not the bride, not the groom, not one of the bridesmaids or family or friends. "I'm not in the habit of crashing weddings," I tell fifteen-year-old Adam Dolaher, whose mom, Sarah, is getting married at the Cyprian Keyes Golf Club in Boylston, Massachusetts, on this late May afternoon. I'm sure he wonders what I'm doing here. "I'm with the birds," I explain.

An hour before the ceremony begins, two white doves in their ornamental cage are already attracting attention. "Oh, my God! Doves! They're so beautiful!" guests cry when they spot them from the patio off the club's music room. Almost no one notices the two crates containing seven other white doves on the lawn outside, even though these birds are the main event of the wedding, as far as I'm concerned.

I've been waiting for weeks to watch a release of white doves. For a while,

White "doves": actually homing pigeons

I found myself in the lugubrious position of hoping for a funeral. Weddings and funerals are the main source of Joe Black's business, Wings of Eternity White Dove Release, and he had promised to take me along to watch his birds fly.

Then, a week ago, Joe got a call from Arni Pinto, the groom at today's wedding, who wanted to surprise his bird-loving bride with a white dove release at the ceremony. Joe couldn't do it that day. He plays bass guitar and sings in a rock band, Ball & Chain. Though he still looks very much the rocker, with his handsome profile, shoulder-length jet hair, and mostly black wardrobe, at age fifty-one, he's scaling back on performances as his dove business ascends. That night, since he had an important gig in Boston, he arranged for his dad to do the release in his place.

Joe couldn't have asked for better backup. He got into the dove business thanks to his dad, Joe Lesieur. Joe Black (he legally changed his last name as he began his career) grew up with aviaries in the backyard. His dad raises and trains champion homing pigeons for races. And in fact, this is what Joe Black's birds are, too: the snow-white birds, symbolizing fidelity, purity, and holy grace, are, like all "doves" released at weddings and funerals, actually trained homing pigeons.

If they were anything else, upon release they would likely meet the fate of eighty unfortunate white birds loosed in Jersey City in 2002. To commemorate the first anniversary of 9/11, the city fathers unwisely selected young squabs bred for their tender breast meat. Never having flown free before, upon release they were lost and confused and immediately flew into spectators, nearby windows, and New York Harbor, where they drowned.

Joe's birds, though, are prepared for this job. "They know what to do," he says—although some of the humans involved do not. Seven people, including the bride and groom, will each get to hold a bird, and on Joe Lesieur's signal, will release them simultaneously into the sky. Some look unsure. One man asks about bird flu. A bridesmaid glares ominously at the groom. "Arni, you are *so* dead meat!" she growls as she is handed a snowy dove. She doesn't like birds, she tells me. It's a good thing she doesn't know what she's holding: the same species that Woody Allen called "rats with wings."

They are all *Columba livia,* the same species of pigeon that crowds city parks, befriends the lonely and the homeless, roosts beneath bridges, and craps on city statues. (Until 2003, the species' official common name was actually rock dove, until the American Ornithologists' Union kicked it out of the doves category and changed its name to rock pigeon. Together the pigeons and doves make up the family Columbidae, with three hundred species, from doves smaller than sparrows to the Victoria crowned pigeon, the size of a turkey.) The same species delivered the results of the first Olympics in 776 B.C. Pigeons brought news of Napoleon's defeat at Waterloo. They saved the Lost Battalion, the U.S. Army's 77th Division trapped behind enemy lines in France's Argonne Forest in World War I. Pigeons fueled the fortune of the famous Rothschild family, who in the early 1800s relied on a network of carrier pigeon lofts throughout Europe to relay information between its financial houses. Even the birds' droppings were once considered so precious that during the sixteenth century, all pigeon shit in the country was declared the property of the English crown, which used it to make gunpowder.

For Joe Black, pigeons were his salvation. Three years ago, his partner died and he sank into a depression. He was sick for a year. He didn't cut his hair. "I looked like Charlie Manson," he told me. Finally he went into the hospital.

Then he met the love of his life, his "earth angel," Karen, and with her encouragement, he started Wings of Eternity. He had seen white dove releases in his travels when he was on tour. "I figured there was an opportunity here," he said. "The wedding industry is a billion-dollar industry. The funeral industry is a billion-dollar industry. And I've got a dad with fifty years' knowledge of birds. I'd be an idiot not to give it a try!"

As Joe puts it, "I'm still involved with rock and record albums—but now I have rock doves, too." The business has brought him and his dad closer together. It's good money. His work is uplifting. What the pigeons have brought him, Joe says, "is like a rebirth."

But this can be a risky business. Not all of Joe's birds always make it back from their engagements. Not long ago, he did a funeral for an army lieutenant who died in a helicopter crash in Iraq. That afternoon he released eighteen birds in Saugus, Massachusetts, a town forty miles away, north and east of the lofts he shares with his dad in Southborough. By night, only a few had made it home. Luckily, all his birds are banded, so when they get lost Joe can be found. "A lot apparently went north instead of southwest. For weeks I got calls from everywhere to come pick up my birds. Birds were all over the place! When they got back they were weak and disoriented. They didn't recover for weeks. Some never did. I wonder what happened out there."

Many things can go wrong. "They're wild animals," Joe tells me in his frank, *r*-less Boston accent. "This isn't the Barnum and Bailey Circus here. You can train 'em good, but sometimes they decide not to cooperate." This happened at a funeral the week before we met. Joe was doing the "Trinity Release." Loosed from their ornamental cage, three birds fly up, symbolizing the Father, Son, and Holy Ghost. As they circle above the gathered mourners, the widow or widower releases a fourth bird, known as the Spirit Bird, who symbolizes the soul of the departed. The Trinity Birds wait for the Spirit to join them, and then all wing off together to heaven. "It's a beautiful thing," says Joe, when all goes as planned.

On that day, things did not go as planned. "The other birds went up and were circling," Joe tells me, "and I gave the widow my smartest, best bird, so I thought. He knows just what to do.

"So she releases the last one," Joe says, " . . . and he plops on top of the casket and just looks at me."

Three times, the bird did the same thing: he jumped from the widow's hand and landed on the casket. "People were starting to murmur," says Joe. "'*Bill's spirit doesn't want leave!*' It was getting eerie. I was so appalled!"

At Joe's advice, the widow finally gave the bird a gentle toss. Finally the pigeon took off.

What had been the matter? I wondered. Did the Spirit Bird perceive some danger in the sky the others didn't see? Was the bird feeling ill? Had some object in the crowd caught the pigeon's eye that held it so spellbound it wouldn't take off? Joe considered this for a moment. I waited to hear some gem of pigeon behavior knowledge. "You know something?" he said thoughtfully. "The guy upstairs works in mysterious ways. Who knows? Perhaps those folks were right. Maybe the spirit didn't want to leave. But boy, I hope *that* doesn't happen again."

Weddings offer their own unique opportunities for disaster. The problems are not what you might expect. Joe has never seen a bride's gown soiled with pigeon droppings (it's unlikely, since the birds, like racing pigeons, aren't fed the night before—it helps ensure a quicker, safer flight home). His birds don't bite—bridesmaids aren't savaged by killer doves. But Joe has seen his share of Bridezillas. "When you start out, they're nice," he says, "but as the day gets closer, they get nervous. They change things. And then you find out the place for the release isn't where they told you it was." Location is crucial to a successful release of a beautiful, swirling white flock. You want wide-open spaces above and in front of the birds. But if the wedding party changes the release location at the last minute, Joe explains—even just a little to the left or to the right, or back behind the reception hall instead of in front—things can go haywire. If there are trees in the way, the pigeons could crash into them, or fly up and land in the branches, and "it won't look right at all."

Timing is also critical. Weddings that start late can be disastrous. "The five thirty release gets delayed till seven, or seven thirty—and you can't let them out that late or they'll be flying through night predators," Joe says. "They start getting nervous when it gets dark. They're timid little animals, they're not eagle warriors."

Happily, tonight's wedding is on time. And Sarah Dolaher is not among the Bridezillas. She speaks sweetly to the gentle white bird in her hands. "Hello, dumpling, it's okay, it's okay," she says. She's thrilled with her new husband's sweet surprise; she beams at the dove in her hand, her expression one of wonder and delight. Her only concern is this: "Do they know where they're going?" she asks Joe's dad. "Are they going to be okay?"

Every bird in the sky has to contend with the same hazards: Hawks. Owls. Winds. Rain. Exhaustion. Angry farmers with guns; nasty kids with slingshots. "Anything can happen," as Joe had told me earlier.

But the wonder of it all is this: most of the time, they come home. On this cloudy spring evening, the seven people in the wedding party gently toss the birds into the air. The assembled guests gasp as one: "Ah!" The white birds circle above like a blessing. "They're getting their bearings," Joe Lesieur tells me. The guests murmur for minutes: how beautiful they are!

But as most of the people go inside for a drink, they miss the real miracle of the evening: released from a site they have never before seen, once they wheel above, mass into a flock, and fly like a prayer into heaven—they return to their loft fifteen miles away in Southborough. Within half an hour—just minutes after we have driven back in our car—they come home.

In the honking V's of fall geese, in the "kettles" of south-flying hawks passing over our mountains, in the loose flocks of songbirds who fly invisibly through the night, wings write the unceasing marvel of migration in the skies. Many mammals migrate, from whales to wildebeest, as do some fish, reptiles, and insects, but it's birds who most capture our hearts. These famously fragile creatures undertake some of the longest journeys on the planet. The distance record holder was the black and white Arctic tern, with its 22,000-mile yearly journey between polar ice caps, but in fall 2006, a team of researchers published news of sooty shearwaters captured in their New Zealand breeding burrows and outfitted with satellite tracking devices. Flying in a giant figure eight over the Pacific basin, they journey 39,000 miles a year. (The birds can also dive beneath the ocean's surface, searching for squid, to 225 feet.) In 2007, an even more astonishing record was established by a bar-tailed godwit. Satellite tracking allowed researchers to follow a female shorebird who flew 7,145 miles *nonstop* from Alaska to New Zealand. In nine days, she crossed the vast Pacific, without a single meal, rest, or drink.

But to me—a woman who cannot drive to Boston without getting unintentionally and yet repeatedly sucked into the Callahan Tunnel leading to Logan Airport—the greatest wonder of bird migration is not the staggering distances flown. It's not that they don't get eaten or injured or exhausted along the way. It's that they don't get *lost*. In some species, such as white-crowned sparrows, the young of the year do not even make their first migration in the company of experienced adults. Yet somehow, without ever having traveled their route, without the benefit of experienced guides, they find their way across a continent. How do they know where to go?

Sometimes I get lost driving to familiar friends' houses. I even get other people lost when they are traveling with me. (Once, trying to return to New Hampshire from a trip to Massachusetts, my falconry-friend Selinda and I had driven for hours before we noticed a sign announcing we had driven south instead of north and were entering the state of Connecticut. We had been embroiled in a discussion of which was smellier: turkey shit or chicken shit.) I trace my problem to growing up as an army brat. On an army base, almost everything looks the same. (The only aspects of the landscape that consistently attracted my attention were usually birds—unfortunately, bad landmarks, because they fly away.) Even after my father retired from the army, we still moved so often I never seemed to get my bearings; though I longed for one, I never had a sense of home. Even now, settled into a house and community where I'll live for the rest of my life, being lost still evokes a kind of panic in me—a panic linked to not being able to find home.

But pigeons can. Though many other species have been studied to probe the mystery of how birds find their way, the species to which most scientists turn for answers is not itself a migrant. *Columba livia* is so ardently wedded to one place that the species has stood for centuries as a symbol of home and fidelity. This pigeon's sense of direction is so refined, and it is so tolerant of human company, that its navigational abilities have been better documented than those of any other bird. Unlike me, they can almost always find home. That is why I had sought the company of homing pigeons, and of the people who know them best. But this is easier said than done. I was very lucky to be able to meet Joe Lesieur and Joe Black. I had gained access to them, and thus entrée to the little-known world of racing homing pigeons, in a somewhat clandestine manner, with the help of a friend of a colleague whom I had never met before.

* * *

We'll call him P. He doesn't want me to use his name. I met him through Steve Bodio, the author of wonderful books on falconry and pigeon racing, with whom I have corresponded on natural history subjects for more than a decade. Steve's never met P., either, but they have corresponded on pigeons. A botanist, farmer, and professional flower grower, P. has put racing on hold for now but is an avid breeder; at Steve's urging, he graciously agrees to give me a tour of his pigeon loft in the Boston suburbs and to introduce me to one of his pigeon mentors. There's only one caveat: "Nobody must know," he tells me, "that I have pigeons."

Why on earth not?

"People aren't in love with pigeons," P. tells me. "Farmers see hoards of them gorging on their grain. Many people see them as flying rats. That's the general perception you have to rise above. And if you have to rise above that general perception, you have an underground mentality. You sometimes have to fly under the radar with pigeons."

P. has never had trouble with his neighbors; the local kids know they're welcome to visit his handsome, gentle birds in their lofts. But, he said, "you'll notice my loft isn't visible from the street." That's on purpose.

Pigeons have an undeserved reputation as filthy, disease-carrying vermin. A single drive-by who took exception to his loft—though it is perfectly legal—could make trouble for him and his birds, he said. What sort of trouble, I wondered? When P. was a kid he kept pigeons as pets. One day he came home from school and discovered that his mother's boyfriend had bulldozed the building, with the pigeons inside. They were crushed alive.

That act of cold-blooded cruelty didn't kill his love for pigeons—racing them, breeding them, watching them. "Once homing pigeons are in you, they're in you," he says. He takes me out back to meet his birds.

I find them cooing and prancing, courting and spinning in the April sunshine filtering through the predator-proof wire mesh of a munchkin-sized second-floor screened porch attached to the neat white-painted coop. One pigeon catches my eye immediately. He's fat, strutting, and proud, with a white collar and iridescent green and purple head. "He's a magnificent showman," P. says. "He's Daddy's big boy." The bird is a Modena, an Italian breed known for its curves: its round erect head, full round breast, and short high tail, giving the creature a pleasing bowl-like shape. This one is not a racer. His job, when the pigeons are flying home in a race, is to act as what is known as a "dropper." He is so charismatic that his very presence acts as a beacon to the racers returning home, inducing them to drop directly into the loft to join him, rather than first circling around in the air, then perching on trees or landing on nearby rooftops and pacing in circles. Since actually entering the home loft is the finish line to a pigeon race, a good dropper can shave valuable seconds off a bird's time.

In his coop, P. introduces me to some of the different pigeon breeds. More than eight hundred breeds exist; pigeons are rivaled only by dogs and perhaps goldfish in the dazzling number of forms a single species can assume. They are so malleable to breeding experiments that Darwin kept them to study the fundamentals of genetics. P. shows me his German Beauty Homers, who have graceful, swan-like heads. These are descendants of the birds Germans used as carrier pigeons in World War I, he tells me. He shows me some Clausson Grizzled pigeons, a breed so tough "you throw one up into a hurricane, they come home." Another, called a Bagdad, is prized for its high flights. He points out two birds who are half Modena, half Tippler. That cross wasn't his idea. They were born of two big birds with brown wings and grey tails, pigeons who are now cooing to each other and tending two eggs in a new nest. He had bought them planning to breed each to other birds. But "they fell in love coming home in the shipping crate," he explained. "I'd love to mate them with someone else, but I'm not a barbarian. So they stay together." Because pigeons often mate for life, their marriage could last for more than eighteen years.

Pigeon racers use the bond between couples to induce racing pigeons to fly home faster. In what's called the "widowhood system," the fancier separates cocks from hens but allows the cocks to see their wives before a race. Sometimes they'll let their racing cocks see the hens with other cocks courting them. Sometimes they'll fly hens when the pair is incubating eggs about to hatch. "They'll fly like mad to get back to those eggs," a racer later tells me, "just like a regular woman."

At P.'s loft, ten pairs of pigeons are raising chicks right now in cozy cupboards built into the coop's indoor walls. Typically pigeons lay two eggs in each clutch. P. will have thirty new birds by the end of the month, he tells me proudly. Several chicks, known as squabs, have hatched already.

"Want to hold a baby?" he offers. Of course I do! P. reaches into a nest box for a chick who is eleven days old. I cup my hands. To my astonishment, he hands me a miniature dodo.

"This is right out of the pages of *Alice in Wonderland*!" I cry. I grew up imprinted on the version with Sir John Tenniel's illustrations, and the Dodo was one of my favorite characters: to a child, the gentlemanly bird was as sweet as he was brilliant, announcing to Alice and her friends that all should join in a "Caucus-race" and then pronouncing, "*Everybody* has won, and *all* must have prizes." I remember my sorrow when I learned that humans had wiped the real dodos off the face of the earth—a travesty worse than the extinction of the dinosaurs! But now I am holding one of those drawings come to life: the same naked head, the same crooked, downturned beak, the same reduced wings and swollen, clumsy body. (Adult dodos' fat bellies wobbled and scraped the ground when they tried to escape from hunters, according to explorers' accounts.) P. reminds me that the dodo was, in fact, a pigeon—as was the passenger pigeon, the most abundant species of bird ever to have lived. And both, of course, are extinct.

The pathetic history of its cousins makes the baby in my hands seem even more vulnerable. Like human infants, baby pigeons are born naked, helpless, and ugly. Even worse, they're blind. They remain in the nest for up to two months before fledging and require extraordinarily dedicated care. Both parents incubate the eggs—the cock usually takes the day shift, the hen, the night—for eighteen days. Both parents feed the babies and keep them warm. The pair's devotion even exceeds those of humans in at least one important respect: *both* parents actually make milk from their own bodies to feed their young. Such a feat is extremely rare in mammals—certain species of male bats can nurse their babies, and there are reports from time to time of human males who do as well. But a *bird* nursing babies? The only ones to do so are pigeons and flamingos.

Though it doesn't come from glands in the breast, "pigeon milk" is chemically similar to mammalian breast milk. A secretion of the adult crop, the first part of the bird's stomach, pigeon milk looks like cottage cheese. It forms the complete diet of the nestlings for the first six days of life and thereafter is supplemented by other regurgitated foods, in the manner of most birds. Both sexes produce pigeon milk in response to the secretion of the pituitary hormone prolactin—the same hormone that controls the production of milk in nursing women and other mammals. The hormone starts flowing while the parents are incubating eggs. In pigeons, sometimes just watching another pigeon incubate or feed babies can induce the production of milk. It reminded me of a friend who had recently had a baby herself; she was shocked when her husband took her out to dinner, and at the sound of a child crying somewhere in the room, she found her own blouse soaked with milk.

"I love every bird I raise," P. tells me. That's already obvious. Several times, I have noticed that after catching a pigeon to show me, before he releases it from his strong, callused hands, P. kisses it on the head. "They're magnificent," he says. "Yes, they're strong and tough. There are some racing guys who tell you they're warriors. But to me, these are creatures that just want to come home. I've always found that endearing."

When he and his wife bought this property—the site of a derelict barn that had been condemned by the city—he disappeared for four days. "Where were you?" his wife asked. He had been working on the new property, he answered. "But I'm building a pigeon loft first," he told her. "Then I'll build our house." P., who ran away from his dysfunctional home at age thirteen, has found a happy home at last, in the company of birds to whom family and homecoming is everything.

Pigeons will do just about anything to get home. They'll fly through rain, wind, hawks, and even bullets. In World War I, to return to the military lofts where they had been raised, Europe's carrier pigeon corps of winged warriors flew through artillery, poison gas, and rifle fire. They made it to their destination 98 percent of the time.

The most famous of them was a pigeon named Cher Ami—Dear Friend—who was as well known to schoolchildren of the 1920s and '30s as any human hero of World War I. One of six hundred birds in the U.S. Army Signal Corps in France, Cher Ami had already delivered eleven important messages within the American sector at Verdun, France, when he was sent up on his last and most important mission on October 4, 1918. The five hundred men of the 77th Infantry Division were trapped behind enemy lines in the Argonne Forest. Within two days, the Germans had killed or wounded half of them. American artillery tried to protect the "Lost Battalion"—but, unsure of the battalion's location, was mistakenly shelling them instead. Fired on by friend and foe, the Americans were out of food, ammunition—and just about out of carrier pigeons. By the third day, the commander had sent up three, and all had been shot down. Cher Ami was the battalion's last chance.

Within seconds of release, Cher Ami was shot in the breast and skull. But even with one eye gone and his left leg shattered, the bloodied bird flew twenty-five miles to reach his home coop in twenty-five minutes. Just hours after the American Signal Corps retrieved the canister holding his message from the shredded ligaments of what remained of the pigeon's left leg, the Lost Battalion was rescued. Though army surgeons worked hard to heal the pigeon's wounds (and soldiers later fashioned him a tiny wooden leg), Cher Ami died of his injuries within a year. But his legacy lives on: he is stuffed and mounted in the Smithsonian's National Museum of American History in Washington, D.C.

Pigeons were essential to the military in World War II as well, even though the use of radio, telephone, and telegraph was widespread, because lines can be cut and messages intercepted. Some were trained to fly to mobile lofts. Some would fly at night. They flew encumbered with miniature cameras to take reconnaissance photos of the land over which they passed. They were released from submarines and parachuted from airplanes into occupied Europe. They flew in the jungles of Burma and the deserts of North Africa. A pigeon named Gustav fought fierce headwinds to cross the English Channel to deliver the first news of D-Day landings. (The heroic bird was later ignominiously killed when a caretaker cleaning his loft mistakenly stepped on him.) Another, named GI Joe, saved more than a thousand British soldiers who almost certainly would otherwise have been mistakenly killed by their American allies. A British brigade had attacked and won back the Italian city of Colvi Vecchia from the German enemy; but because no radio was working, they were unable to reach their American allies, who were planning a raid on the same city. GI Joe arrived at the American air base just before the pilots were to take off. A Scottish pigeon, Winkie, escaped from her cage aboard a British bomber after the plane fell into the North Sea. Though covered with oil, she flew 129 miles, at night, to deliver the crew's coordinates to rescue teams on shore in Scotland.

Winkie became one of the first to receive a new medal to honor animals displaying "conspicuous gallantry and devotion to duty whilst serving with the British Commonwealth armed forces" or civil emergency services. Created by the founder of the People's Dispensary for Sick Animals, Maria Dickin, the Dickin Medal is today recognized as the animal analog of the Victoria Cross, the highest and most famous wartime military decoration of the Western world. Only sixty-two animals have been so honored since the medal was created in 1943 (five were recently awarded to dogs serving in the former Yugoslavia, in America after the September 11 attacks, and in Iraq and Afghanistan). Thirty-two of them were awarded to pigeons.

Our kind is lucky that these brave, strong, devoted birds are seedeaters. That's why this particular species of pigeon has chosen to live with us, probably ever since humans began to grow grain. To further induce rock pigeons to stay nearby, people built pigeon houses, or dovecotes, as convenient alternatives to the birds' natural cliff dwellings, with clay pots inside to accommodate nests of sticks, twigs, and grasses. Pigeon keeping spread into Europe with the expansion of the Roman Empire. From the Renaissance to the early 1800s, pigeons oversaw the flowering of European culture from special open towers with hipped roofs built just for them. Aristocrats adorned their properties with elaborate dovecotes resembling classical temples or medieval towers.

The birds well deserved such luxury, for they had already been serving their human hosts for millennia. Originally found wild from the cliffs of coastal Scotland through the deserts of Central Asia and North Africa into India, pigeons may have been first domesticated for meat, for pets, or possibly for the rich fertilizer provided by their droppings (much later the stuff was prized as the only source of saltpeter for gunpowder). But early civilizations quickly realized pigeons' potential as messengers. Pigeons carried ancient Egyptians' news along the Nile when the life-giving floodwaters returned. The writers of the Old Testament knew pigeons were trustworthy. Noah's raven never returned, but his dove—almost certainly a pigeon—did, and with lifesaving news. King Solomon and Cyrus the Great used pigeons to carry critical messages. The ancient Greeks learned to do so as well: a human runner took most of the day to bring news of the Persian defeat at Marathon to Athens, only twenty-six miles away, and then died of exhaustion. A pigeon could have made the trip in less than an hour. Even the world's greatest horsemen conceded that, when it came to delivering messages, their steeds were outmatched by pigeons. Destinations that take a horse and rider days to reach, a pigeon can fly to in hours. Mongolian emperor Kublai Khan knew this and established a pigeon post that covered one-sixth of the globe.

In his book, *Pigeons,* journalist Andrew Blechman compares the athletic abilities of horses and pigeons: a horse runs thirty-five miles an hour over a one-mile race. A pigeon can fly more than five hundred miles in one day at speeds over sixty miles per hour—and the bird performs the feat fueled by just an ounce of seed eaten the day before, and without stopping for a sip of water.

Pigeons would even beat the iron horse. One of the world's most famous and enduring news agencies, Reuters, began its European business in 1850 with homing pigeons, flying from Aachen in Germany to Brussels in Belgium. The telegraph service had gaps and the rail was too slow. The pigeons, flying seventy-six miles in two hours, beat the railway by four hours.

It's easy to see why racing these avian athletes against one another took hold as a popular sport. Belgian breeders are credited with the creation of modern pigeon racing, starting in the early 1800s. Races are won by seconds, and wanting your own bird to win, my friend Steve Bodio wrote in his book *Aloft,* feels "as real as hunger." It's a storied passion in which P. is proud to take part.

He loves the birds, and he loves the human characters who gather around them in racing pigeon clubs. Many fanciers are wealthy, but wealth doesn't guarantee a win. P. tells me about one guy in his local pigeon club, a man so ancient and sickly he could barely walk. When he moved into a retirement home, he moved his pigeons from his backyard loft to an abandoned school bus on the pigeon club's property. The man would get over there to feed the pigeons when he could; others in the club would feed them the rest of the time. "They were almost left to fend for themselves," P. remembers. "And he would enter these almost feral pigeons into these five-hundred-mile races— and they would win! Why? They were tough."

"Pigeons are like the red-headed stepchildren of the world," P. tells me, "but they have more stamina than almost anything else on earth."

How best to tap that stamina is the subject of closely guarded secrets among pigeon racers: they know a formula for "tea" to make the birds stronger; a way to bring on or stave off the molting of feathers with artificial lights; how to predict a winner by the configuration of the pigeon's eye. (I actually read an entire book on this but still don't fully understand it.) Each fancier has different, often elaborate theories about everything from training, breeding, and feeding to the shape and consistency of healthy droppings. Who would be willing to share this with an author, especially someone they had never met before?

Happily for me, P. introduced me to Joe Lesieur.

Even though P. is forty-nine, he still does not call his friend by his first name. When P. started racing pigeons, Joe Lesieur was a mentor. P. chose wisely and still respects him so much that he still calls him "Mister Lesieur." So I do, too.

Mr. Lesieur is a former engineer, now a custodian for a local school, a trim man of seventy-three with blue eyes and neat steel-colored hair parted to the side. He's owned pigeons since his own dad brought him two pair as pets when he was nine and has been racing them since he was a teen. "It took me fifty years to learn what I know," Mr. Lesieur tells me, "and I still don't know anything!" But he is too modest. Pigeon fanciers ship their six- to eight-week-old pigeons from across the nation to him to train for the American Racing Pigeon Union convention in New England, three days of races drawing birds from one hundred lofts to compete for a top purse of thirty thousand dollars. Birds he owns and trains have won numerous prizes. In one race, his birds won first, second, third, fourth, fifth, sixth, and seventh places. The win for which he's perhaps best known was a race in 1999, the John "Papa" Orsie Memorial Race, one of the largest Young Bird races in New England. One hundred birds from twenty-four lofts competed in that race; one of Joe Lesieur's came in first. His photo holding the winning pigeon, a hen hatched that year, was featured on the cover of *Racing Pigeon Bulletin*. "There's guys who would kill for that," he told me later. "But to me, it's nothing. The important thing is the birds fly."

Some of the birds Mr. Lesieur trains cost thousands of dollars—and they can be well worth the investment. Some races offer handsome purses as prizes; others, none at all. "But this isn't where the big money is," P. tells me. "If your pigeon is a real winner, he can be worth a great deal as a breeder, just like a Thoroughbred racehorse." (Highest price on record at this writing for a stud racing pigeon: $132,517.) "And you can gamble on your bird," notes P., "knowing you have a good chance of winning—just like you bet on the horses."

But nothing is less like a horse race than a pigeon race, as I would soon find out.

"... *And they're off! It's Join in the Dance racing for the lead, Musket Man has some early speed, and farther back on the rail is Atomic Rain* ..." On the first Saturday in May, my husband and I, along with 16 million others, are watching the Kentucky Derby on TV, our hearts keeping time with the announcer's frenzied narration and the horses' pounding hooves. *"At the back of the pack, moving up now, is Hold Me Back* ... *Chocolate Candy is coming alive* ... *and at the top of the stretch, Musket Man is coming hard, now toward the inside is Mine That Bird* ..."

At the finish line, the announcer is beside himself with excitement: "A SPECTACULAR UPSET!" A small, unheralded gelding, a bargain basement buy at $9,500, a 50-to-1 long shot, *"Mine That Bird has won the Kentucky Derby!"*

The following Saturday, I would taste the excitement of a very different race. Mr. Lesieur and I would spend most of a day sitting in lawn chairs facing his loft, waiting for a pigeon to land.

A pigeon race, observes Steve Bodio, is "the strangest of all kinds of racing." There are no crowds. Only one person witnesses the starting line. The race itself is invisible. Every contestant flies a slightly different course. And birds from every loft cross a different finish line.

On this breezy May day, the starting line is Ilion, New York, about two hundred miles away. This competition is confined to "Old Birds" one year old and up. Seven hundred and thirty-nine pigeons from sixty-two lofts belonging to nine pigeon clubs are competing. They are freed, all at once, with the flip of a single lever, from rectangular wire crates loaded into the huge truck that had driven them to the liberation point the night before. Sometimes the pigeons are liberated as early as 6:00 a.m. But today, because the weather was rainy earlier in Ilion, the race doesn't start till 8:00. At release in Ilion, we learn, it was 45 degrees Fahrenheit and cloudy with a light south wind. In Southborough, when I arrive around 10:30, the sky alternates spitting rain with bright sun, and it's cool and breezy enough that I zip up my polar fleece.

"Birds could get here in three hours forty-five minutes, maybe four," Mr. Lesieur predicts as we settle into our chairs, about twenty yards from the loft. "We should start looking for them around eleven thirty—though we probably won't see any till noon." But you never know. "I've seen days I figure the birds'll be here in three hours—and you look up, and there they are," he says. And other days, he says, the birds don't come home for hours and hours. The last of the group may straggle home the next day. Or never return at all.

For someone about to spend many hours sitting in a lawn chair, I'm both nervous and excited. Since P. first introduced us, I've come to visit Mr. Lesieur several times and found this modest, wry gentleman to be everything P. promised: open, kind, and knowledgeable. His immaculate, spacious lofts smell of lemon disinfectant and his birds are in superb shape. He and his wife love animals: two Chihuahuas sleep in the bed with them each night, one of whom they have nursed to good-natured, robust health from a sickly, fearful, biting shelter waif. Two parakeets chirp cheerfully from a cage in the small dining room. He greets the wild rabbits who roam the backyard.

Mr. Lesieur and his wife, Esther, are quietly but breathtakingly generous. One day when he learned P. and I were coming over, Mr. Lesieur had bought hot dogs and hamburgers to fix for us on the grill. When he discovered I don't eat meat, he disappeared into the house, and within minutes, his wife, only weeks after major abdominal surgery, appeared with a luscious salad.

Warm welcomes are routine in this big, close-knit family. Mr. Lesieur was one of eight kids, his wife one of eighteen kids, and their house is always full of people: friends, grandchildren, cousins, aunts, sisters, musicians with Joe Black's band. The family hosts reunions for 150 people at the drop of a hat. The couple fell in love as teenagers, when she was a pretty majorette in high school; they have lived in this two-family house for fifty years, one town over from where both grew up.

In their devotion to home and family, they remind me of pigeons. Mr. Lesieur claims not to care too much for his individual birds ("The one you like is the one you'll lose," he warns his son Joe). But one day he admitted: "What steals your heart is, they come back. They want to come home." I have come to like Mr. Lesieur, his family, and his birds very much; I want all his pigeons to come home; and I want one of his birds to win.

And one very well might. He didn't fly his birds much last year, because his wife had been ill; he missed last week's race because of the Mother's Day weekend. But among the fifteen cocks he's flying today are several proven, experienced winners. He's been taking them out every decent day for practice flights, loading them into crates and driving them for release sometimes ten miles away, sometimes fifty. One afternoon when I visited him, it was too hot to fly, up in the nineties, but he had already gotten up that morning at 4:40 to fly them when the day was still cool. The week before, he took them out one hundred miles, letting them go in small groups so they could race one another home. "They've been making good time," he tells me. Today's southerly winds will push birds returning to the concourse's more northerly lofts first; they will be at an advantage. But if his birds "are on the ball," he says, "I could win the south section."

How do you proclaim a winner when each loft is at a different distance from the starting line? The first bird to make it home isn't necessarily the winner. The winning time is calculated in yards per minute. The Lesieurs' J & J Loft is 182.4 air miles from the Ilion starting line, as determined by the Global Positioning System. When a bird lands on a rubber pad in front of the loft door, a scanner records the number of the band on the leg, like an item at the grocery store. The bird's number, as well as the exact time of landing, is recorded on a special computerized clock. Divide the loft's distance from the starting line by the time in minutes it took the bird to return and multiply by the number of yards in a mile to yield yards per minute. Only comparing this time with birds of other lofts—which is done at the pigeon club later that night when everyone brings in his calibrated clock—determines the winner.

Last week's race, Mr. Lesieur tells me, was a fast one. Six hundred and fifty-seven birds from fifty-six lofts were released at 8:30 a.m. There was a west wind, pushing the birds east—and the winning bird came in at 1,894.345 yards per minute. "That's about ninety miles per hour," Mr. Lesieur says.

"But that's not going to happen today." Today the birds may be flying through rain, and the winds are against them.

"The wind controls the whole race," he tells me. "When we let them up in the west, and fly west to east, an easterly wind is a headwind. Now with a tail wind, we could fly a hundred miles in an hour and forty-five minutes. But with an easterly wind, the same course could take four hours."

Because of the location of his loft relative to the other competitors', his birds do best with a northwest/southwest wind, he explained. "But if it's a decent day and the birds aren't going to get hurt, I'm going to fly them. It doesn't matter if I win. I just want my birds to fly."

So we settle into our chairs, facing north and slightly east, looking at the landing platform on the loft as if it's a TV set in a living room, and wait. Most people will never notice, but today racing pigeons are everywhere. His club is part of the Greater Boston Concourse, and it's not the only group with a race today. Within a 150-mile radius of where we sit, from Fall River, Massachusetts, to Cape Cod to New Hampshire, there must be two thousand homing pigeons out today, all trying to fly home.

What goes on in the skies above us, unseen? Only one person witnesses the release: the liberator, which in this case is Art MacKinnon, whom I met one night at the pigeon club. When he turns the lever that opens all the cages at the release site, "they shoot out like water out of a hose," he says. They usually circle as a group and start out flying as one big flock, even though many of the pigeons have never met before; they know there is safety in numbers. To get home, they must fly east. "If they see the sun, they go toward that. But if they're young birds, they could go any direction," MacKinnon reports.

And then what?

Dr. Charles Walcott of Cornell University has been studying homing pigeons since 1962 trying to answer that question. During the high point of his studies he worked with two thousand homing pigeons in his test group. He would release them from different sites, follow them with binoculars as they disappeared to get their "vanishing bearings," and, as many had radio transmitters attached to their backs, track them from a small plane that he flew.

He took them from their lofts in Ithaca, New York, for release in Framingham, Massachusetts. A major highway goes straight from Framingham almost all the way to the lofts at the Cornell bio labs; the road is an excellent landmark easily seen from the sky. "If I were a pigeon," he said, "I'd take that route—but they didn't." Why? He still doesn't know.

Another mystery: Even though the skies free them to do so, the Cornell study pigeons don't fly straight home ("as the crow flies") but usually take twists and turns, even when they don't land to rest or feed. And even when they are released from the same location, they don't take the same route home each time. This isn't the case with other pigeons in other studies, however: a ten-year study at Oxford University found that their pigeons *did* take the same routes. The Oxford pigeons relied so heavily on roads and freeways that they would sometimes even switch directions at freeway junctions.

How can different institutions studying the same species come to such different conclusions? In his studies, Walcott told me, "The thing we were impressed most by is that they're such individuals." Even though birds like to fly in a flock, there are mavericks. One pigeon flew from Fitchburg, Massachusetts, to Mount Wachusett, and then to its loft at Harvard. Then, released from Worcester, Massachusetts, the same bird flew directly back to Mount Wachusett, even though this was a detour, then turned right to Lincoln, Massachusetts, and then finally headed to Cambridge. One bird he remembers especially was Blue 38. "This pigeon had a very good navigational ability but couldn't find the loft," he recalled. Released from almost anywhere, the bird flew unerringly back to Ithaca—but never to the coop. The pigeon would find someone out gardening and land on their lawn. The person would see the band on the leg, call the 800 number for the pigeon lab, and send the bird home in a taxi.

That each bird was so different, Walcott said, "wasn't a surprise to the pigeon folk." He often worked with homing pigeon clubs and found this to be true: Trainers already knew, quite well, this fact that the scientists, in the thrall of their statistics, took so long to discover.

*   *   *

"I would say we'll get birds in the next hour or so," says Mr. Lesieur. It's a little after 11:00 a.m. "But I don't know what they could run into," he continues. "Now the wind is coming from the west. If that keeps up, they'll fly in over our heads. They might be happy to get home. They might make one circle and then—boom! If I'm out here, they might not circle. Or they might be tired . . ." But he dismisses this thought. "They seemed strong . . ."

Meanwhile, Joe Black, who lives in an apartment on his folks' property, decides to take his whites out for a practice flight. They need training, too, though their flights are far less rigorous than the racers'. He urges twenty of them from their coop into a carrying crate, packs them into the car, and drives off to release them at a nearby park, ten miles away.

At 11:35 we see Joe's whites swooping over us, a live, swirling, snowy cloud. The brilliant white pigeons circle again and again. It's obvious they take joy in their flight. One bird takes the lead, then relinquishes it to another, then another, and another. How do they avoid crashing into one another? Researchers studying huge flocks of birds like the million starlings who mass over the coastal wetlands of Denmark discovered that birds in flight do bump into one another, but seldom with disastrous results. Individual birds don't distribute themselves evenly when flying in flocks. Using images gathered from high-speed stereoscopic photography and computer software borrowed from the field of statistical mechanics (which seeks to explain how materials behave by examining their molecular structure), physicists mapping these flocks find that like drivers on a highway, individual birds don't seem to mind neighbors nearby on the sides, above or below—but they like a good deal of space behind and in front. Each bird in the flying flock, it appears, keeps track of the position of six or seven birds nearby. To do so, they watch, they listen, and they probably also pay attention to the feel of the air moving through the wings of their neighbors.

We hear the clap of wings as Joe's white pigeons settle on the roof of the house. But they don't want to go inside the loft yet.

Joe returns in his car five minutes later. Still none of his birds has gone back in the coop. "C'mon! Ca'mon!" Joe calls, then whistles. His dad calls

"C'mon, guys!" in a sweet voice. But nobody's going in. They're all perching on the roof. Joe shoos them away with a pool scrubber. This way they'll get more exercise and maybe then want to go inside for a drink and a snack. They fly, circling once, twice, thrice. After a fourth circle, one bird lands and walks straight into the coop. The others keep flying. "It's actually better that they fly," Joe's dad tells him gently. "It makes them stronger and stronger."

But Joe is distraught. "This malarkey today is because I *want* them to go in!" he says. Members of his band are coming shortly to rehearse for a concert, and he doesn't want to retire to the studio worried about his pretty white birds sitting outside attracting hawks. Three have already been killed this spring—including his favorite, Dirtman, whose white face and neck were speckled with black. "He was one of my firstborns last year," Joe says sadly. "Always at the front, always brought the flock home. One day I did a job in Winchester. I lost eight birds that day. It was a narrow spot to put them up. A pond behind, a building and trees ahead, and the ceremony started late. I put them up at six thirty. I let twenty go, and only got twelve back out of twenty. That was a bad, bad day."

I'm hoping that today won't be another.

By noon, Mr. Lesieur and I have seen a number of different birds: Two doves land on the lawn and coo; a red-winged blackbird calls "onk-a-REE!" at a moment the sun breaks through the clouds. We watch a vulture soar overhead. "If only we could fly like a bird!" Mr. Lesieur muses. "We'd be up gliding around . . ." A great blue heron sails off toward his favorite pond, just blocks from here. We see chickadees, crows, two pretty orioles . . . But where is a pigeon when you want one?

"I'm willing one to materialize from the sky," I say to Mr. Lesieur.

"I don't watch the air," Mr. Lesieur tells me. "I watch *these* pigeons." Joe's whites are milling in the screened porch area of his coop. Bowing, bobbing, spinning, and cooing, they're courting—which pigeons, unlike most birds, do at all times of the year. They look like elegant brides and grooms in wedding gowns and white tuxes and tails. But they also keep an eye on the sky.

When we were at the wedding, I had noticed that the moment Mr. Lesieur put the two crates of pigeons down on the lawn at the golf club, the birds went quiet and still. They were looking upward, first with one eye, then the other.

"They'll see a pigeon coming before anybody else," Mr. Lesieur says. "They watch everything. You look where they're looking at, and there'll be a bird up so high you can't see it. But *they* do."

Not much gets past a pigeon. They notice details that humans miss: one study found that pigeons could learn to recognize the difference between the painting style of Manet and that of Monet faster than many college students. At one time, the U.S. Coast Guard trained pigeons in helicopters to spot orange life vests at sea; they outperformed human spotters three to one.

So much of this world escapes human notice, and much more is beyond our perception. But it's not beyond a pigeon's.

"When I started this work," researcher Charlie Walcott told me, "pigeons were assumed to have the same capabilities we do. But they have senses we do not. They see ultraviolet light. They hear infrasound, below the threshold of human hearing. They sense tiny changes in barometric pressure. They sense the electromagnetic field of the earth."

Pigeons are blessed with bits of magnetite in the upper part of the beak, which might serve them like a compass needle or help them to identify magnetic landmarks along a learned route. But birds have a second mechanism for gathering and judging information about the earth's magnetic field as well. When light strikes the retina in the eye, it changes the conformation of its molecules. In birds, though, another change also occurs in the presence of a weak magnetic field. This alters the way in which one of the proteins in the retina—rhodopsin, a pigment responsible for night vision as well as blue-green color vision—behaves. By injecting wild-caught garden warblers' eyes and brains with tracers, German researchers have established a direct neural connection between the retina and the portion of the brain involved in compass orientation. The study strongly suggests that birds can actually *see* the magnetic field of the earth.

And this is just one sensory tool birds use as they navigate the earth. They may also use the infrasonic roar of distant ocean waves or wind in far-off mountains as acoustic landmarks. They can use the sun's pathway across the sky as a compass. They can use the angle of polarized light to determine direction. Italian research with pigeons suggests that they rely also upon smell to map their world.

Charlie Walcott feels certain his pigeons use all these cues, and possibly more—along with their proven ability to learn and their excellent memories. "There isn't just one answer" to how homing pigeons find their way "but many answers," he says. Like pilots, birds surely need at least one backup to even the most accurate and reliable navigation system. The sun compass isn't enough, for the sun doesn't always shine. The magnetic sense can be disrupted; strong solar winds distort the field, and iron deposits create magnetic anomalies. Sounds can be masked or warped; scents blocked or swept away. No one system always works. And so many things can go wrong.

It's half past noon. Another crow flies by. A blackbird alights in a tree. Six of Joe's whites have finally gone into the coop; the stragglers circle above. Mr. Lesieur's cell phone battery has gone dead. While I watch for pigeons, Mr. Lesieur goes inside to recharge it, and from the wall phone he calls the club's race secretary, Peter Shalginewicz, over in Brockton. Have *anyone's* birds come home yet?

He emerges from the house with a worrying report: "It looks like it's going to be tough. It might be a five- or six-hour race. You're getting showers and fog and rain and south wind. Everything is against us."

At 12:45 Mr. Lesieur again disappears into the house. I later learn he is eating some sort of meat for lunch and wants to spare me a sight he fears would distress me; he emerges with another salad his wife left for my lunch. "They could have run into heavy rain," he tells me as I munch my lettuce. "Then all the birds go down. They didn't hold off the race today because they knew it would be worse tomorrow. Tomorrow's supposed to be solid rain."

By the time I finish my immense salad, it's nearly one o'clock. "I've never seen pigeons take this long!" Mr. Lesieur says. "It's almost five hours! It's the winds and the weather." We might wait all night before the birds are back.

At 1:18: "I would say we should have had some by now," Mr. Lesieur says. "I wish they could talk. They could tell me all the things I'm doing wrong."

At 1:32, we can't stand it anymore. Mr. Lesieur phones the race secretary again. "Peter? I haven't seen a feather! How many you got? What time did they get in?"

Mr. Lesieur discovers that the first bird in the club came down around noon. "Remember, I said I thought it might take four hours before a bird came down? I wasn't wrong," he tells me, chuckling. "It just wasn't mine."

Any hope of winning this race is lost. The other lofts in the club all have had at least one pigeon return by now. "Everybody's got birds but me," Mr. Lesieur says. "It's only one race. But I've never seen it as bad as this. This is what happens. These birds haven't flown that much for a couple years. They get lazy. But they'll come home . . .

"I'd a thought they'd have done better than this. But we'll ship them next week for the three-hundred-mile race. When they come home today, I'll give them a little rest. What's important is the birds fly, and that they come home. They'll be smarter from this.

"They'll come up over those trees," he predicts. And then he admits, "I'm dying to see one now."

In their efforts to study the tools that pigeons use to find direction and position, scientists have repeatedly tried to thwart them. They've released pigeons with magnets attached to their beaks, coils attached to their heads, and ears plugged with cotton. Pigeons have flown wearing frosted goggles and with nerves to different areas of the brain cut.

And still, no matter what the scientists do to them, at least some of the pigeons still manage to find their way home. Even when one laboratory demonstrates an experiment in which their birds' sense of direction was disabled, another lab performs the same experiment and *their* birds make it home fine.

The discrepancy is apparently not due to genetics; the researchers have tested for that. And it's apparently not due to differences in the lands over which the pigeons fly. It seems instead to have something to do with the pigeons' upbringing at their home loft—essentially, their loft culture.

Charlie Walcott saw this for himself at an unremarkable hilltop, seventy miles west of Ithaca, called Jersey Hill.

Because it has a fire tower, it seemed a good place from which to release homing pigeons and then watch them with binoculars. But every time the Cornell team released pigeons from Jersey Hill, only 10 percent of the birds made it home. "It was a Bermuda Triangle for pigeons," Charlie Walcott told me.

"We couldn't figure out the problem," he said. "We followed in airplanes. We couldn't figure out what was going wrong." So they asked local pigeon racers if they would lend them some of their homing pigeons to release from there.

"From Jersey Hill?" asked the trainers. "Isn't that the place pigeons never come home from?" They had read the scientific papers Walcott had published on Jersey Hill and weren't eager to lose their own birds.

But a weird thing happened. A few brave fanciers lent their precious racing pigeons to the researchers. They were released at Jersey Hill.

They came home fine.

"This must be telling us something very important," said Walcott. "Those pigeons were raised only ten miles from Cornell. It's clear that our pigeons are comparing their experience at Cornell with something that isn't present at Jersey Hill. But we're not smart enough to know what it is."

I wonder: Have Mr. Lesieur's pigeons run into some problem uniquely afflicting them? What's taking them so long?

At 1:50, the sky is dark and ominous. I put on a rain jacket over my polar fleece. "I'm worried," I say.

"They'll come home," says Mr. Lesieur. "They want to see the girls."

"This is stressful," I say.

"It is if you let it bother you," he says. "But I know birds do what they want. When pigeons aren't ready, this is what happens. They've been out a hundred miles before, but not with other birds. They flew a hundred miles fine last week. They just don't know what to do. They don't know when to break from the flock. And if they hit rain, they'll go down."

At 2:22, Mr. Lesieur calls the club secretary again. "Peter's only got three of his fifteen birds in, and they were well-trained birds," he reports. "All the guys got a couple birds, which tells you they hit bad weather. They could have landed on a big building somewhere and just stopped together." If they have done so, at least there is safety in numbers, and the hawks won't be out hunting in the rain.

By 2:30, they're still not here: "This is discouraging!" says Mr. Lesieur. "I've never had them come this bad."

At 3:00, he calls it a day. He doesn't want to sit out back anymore in the crummy weather. It's depressing. The electronic clock will record when the birds come in, and it won't matter anyway because it's obvious his birds haven't won. He hands me his cell phone for me to call my husband to tell him I'm coming home.

But before I finish dialing—wings! A pigeon with markings called "blue bar" because of the dark stripes on each wing sails in from the north, makes one circle above the coop, and touches down. "Go ahead, guy!" urges Mr. Lesieur—and the pigeon, unlike Joe's whites, walks directly into the trap door. We rush to check the clock. It records that Number 106, a three-year-old cock, has arrived at 15:05:35—3:05 p.m. and thirty-five seconds. Mr. Lesieur goes into the loft to examine his bird. Is he all right? Wet? Hurt? Exhausted? "He ain't fat," he says, feeling the breastbone. "He's been on the wing for seven hours."

As we emerge from the coop, we are greeted with the sight of another pigeon. At 15:13.37—3:13 p.m. and thirty-seven seconds—another cock flies in, a three-year-old with a "check" pattern on the wings bearing the number 137, and walks right into the coop. The birds are coming back. They are going to be all right.

Immediately after 137, a third racing pigeon alights on the roof. At a glance, Mr. Lesieur knows this one's not his. "But he's going home," he says, "wherever that may be."

All the pigeons made it back to the loft that afternoon, except for one who returned the next morning, and one, a purplish cock, who never returned at all. Joe thinks maybe he landed at a different loft and that his new owner is delighted to have him. Now it's a week later, and the pigeons are getting ready for another race.

Normally, races are flown on Saturdays. But on Friday, a storm coming out of the southwest promised rain the next day in upstate New York. The forecast for Sunday was better. So the birds will be shipped the three hundred miles to Lyons, New York, on Saturday, with plans for a Sunday race.

I get to the Lesieurs' house the evening before the race. "I think it will be southeast winds," says Mr. Lesieur as we sit in the kitchen eating slices of white cake and drinking coffee, "and if it is—"

"Joe loses," his wife pipes in. She's been hearing this sort of thing for years.

"No, I could win. I could win—OR lose," he says. "You never know about these things." But of course he is imagining how things *might* be: "If it's clear they'll let them up at eight . . . The storm's coming from the southwest, and if it keeps up, they won't be home for five or six hours. Unless the wind turns around and becomes a tail wind . . ."

Either way, he has high hopes for his birds. "They learned something from that race last week. Besides, I'm not doing it to win. I do it to enjoy them."

He also does it to enjoy the company of his fellow fanciers at the Norwood Homing Pigeon Club. From the house in Southborough to the club is a forty-five-minute, crazy-rush-hour, Massachusetts-type drive up Route 9 to the MassPike—one that would surely suck me directly into the Callahan Tunnel and funnel me helplessly to the airport. But fortunately, Mr. Lesieur is driving, and it's a trip he undertakes happily every time he races his birds. He's taking me with him to meet the guys at the club. They gather here the night before each race for the "shipping event," when each loft's clocks are calibrated and the racing birds are transferred to the shipping crates in which they will ride to their starting line.

"These guys are good fliers," he tells me as he drives. "If you can win here, you can win anywhere."

Escaping the MassPike, we pull into the edge of the suburb of Norwood and up to a sizable lot hosting a low white building about the size of a barn, emblazoned "Norwood Racing Pigeon Club." To my astonishment, among the eight club members present on this night are pigeon racers gathered from around the globe.

The first person I meet is James Cheng, a retired businessman from Taiwan, who now lives in Sharon, Massachusetts. His English is still new, but with the help of gestures and a stick that he uses to draw in the dirt, he relays that he has been racing pigeons since he was a child. He would take his pigeons in their basket with him on his bicycle, release them, and they'd be waiting for him when he biked home. In today's Taiwan, however, the sport is only for rich men. Just the bands on the feet of each bird may cost one hundred dollars or more, and they are often made of gold or silver. Purses may reach $3 million in what I read on one website was considered "the richest, most demanding races in the world." There, racing homers have grown so valuable that the pigeons are almost always liberated from ships offshore—for to release them for an overland flight is to risk kidnapping by professionals who hold them for ransom.

Taiwan's national passion for pigeons, I read, "borders on mania." The little leaf-shaped island off the east coast of Asia is only twice the size of New Jersey, but some eighty thousand people are involved in racing, breeding, raising, and training homing pigeons. Taiwan hosts thousands of local races, some featuring the release of twelve thousand birds at a time. Racers may wait two years to join a particularly prestigious pigeon club. Before a race, people will burn joss sticks at temples and pray for their birds' swift, safe return.

Pigeon racing was introduced to Taiwan from Japan at the turn of the last century; the country's obsession with the sport really began only about forty years ago. But in Ireland, where Alan Power comes from, folks have been doing it for centuries. "Some places in Ireland," he tells me in his lilting accent, "every second house has a pigeon loft." Alan is the only member of the club I meet tonight who is not retired; he's twenty-eight, and the older guys are glad to have him. Like most of them, Alan started racing pigeons as a child; when he came to the United States six years ago, one of the first things he did was join the pigeon club; in the new country, racing pigeons makes him feel more at home.

The same was true of Ardi Farhadi, the pigeon club's vice president. Tall, dark, and stately, he grew up in Iran, flying pigeons from the rooftops. One day after he moved to New England in 1974, while driving (on purpose) to Logan, he noticed homing pigeons flying overhead. He remembers thinking then: "Oh good! One day, I will have pigeons again."

But American pigeon racers do several things differently than they do in Iran. For instance, Americans do not perfume their pigeons. In Iran, each loft has its own signature scent—cinnamon, mint, oregano—with which the people anoint their pigeons before each race. "I think scent is very important to pigeons," he tells me.

Many researchers agree. Most birds (with the obvious exception of carrion-seeking New World vultures) don't seem to have an acute sense of smell, but pigeons are different. In Italian experiments, if allowed to smell the air on their trip to a new release site, pigeons can find their home; if their nostrils are blocked or the olfactory nerves cut, they cannot. (Interestingly, Cornell pigeons seemed unaffected.) Such data suggest that the Italian pigeons, at least, map their world with odor: they know to turn left at the sewage treatment plant, or right at the apple orchard.

An intriguing new paper published in spring 2009 suggests that the way scent works might be even more complicated than that. Lisbon-based researchers collaborating with a Virginia Tech colleague divided their pigeons into three groups. One group of young birds was allowed to smell the natural odors along the route to the release site. Another breathed only synthetic bottled air. A third group was exposed to a series of novel scents that had nothing to do with the world outside them: lavender, camellia, eucalyptus, rose, and jasmine scents were piped in, in that order, and added to their breathing supply of synthetic bottled air.

To no one's surprise, upon release, the first group of birds navigated beautifully. The second group, breathing scentless, bottled air, was disoriented. But what astonished the researchers was that the third group—the one exposed to novel, random odors, completely unrelated to the actual landscape—fared as well as the first! Lead author Paulo Jorge of Portugal's national museum of natural history and coauthors suggest a fascinating explanation: rather than providing olfactory landmarks, odors might inspire birds to activate a different kind of navigational system—possibly the same way scent cues like Proust's madeleines trigger human memory.

I want to ask Mr. Farhadi what he thinks about this, but there's no time. Promptly at 7:00—the truck usually shows up around 7:30—the members of the club start scanning the bands on the feet of the birds into the club computer, officially registering each bird. To do this most efficiently, the men form a sort of assembly line. One man takes each bird out of its crate and hands it to the next man, who scans the bands and then hands them off to a third fellow, who puts each in his or her traveling crate. (The hens and cocks are separated to avoid romantic distractions.) The men hold the birds gently, with the feet grasped firmly between the second and third fingers of the right hand and the breast cradled by the left palm. The feet are up and the head is down so the scanning can be done with the least stress to the birds possible. As the men work, they comment on one another's pigeons, evaluating their health with their hands: "Yours feel good, Joe," Mr. Farhadi tells Mr. Lesieur.

Racing homer in the hands of its trainer

"Well, they gotta come better than last week," he replies.

Dave Gage, who has brought fifteen birds to fly in tomorrow's race, predicts the winds will come from the west or northwest. "Sometimes they come up pretty quick. There could be gusty tail winds . . ."

I step back for a moment and admire the eight men before me. I've heard Mr. Lesieur speak of Dave Gage before; he admires his birds. ("If mine are on the ball, they'll follow his!") Bob Bibbo of Leominster is a contender to win tomorrow's race, too. Al Casella, a retired widower from Weston, has been racing pigeons for sixty-five years. In this room, gathered from four countries on three continents, are nearly four hundred combined years of racing experience and an inestimable store of passion. Peter Shalginewicz, the club secretary, is wearing a gas mask as he scans the birds. I ask him why he's wearing it. He removes it for a moment to explain: he's violently allergic to birds. He puts it back on and continues scanning the pigeons.

The passion is catching. I wake on Sunday wondering about the weather. I look out the window: Damn! It's raining.

But I'm in New Hampshire; the pigeons are in New York and flying to Massachusetts. I check the Internet. It's raining in Lyons, too. At 8:00 a.m., I phone Mr. Lesieur. "It's pouring down here," he reports. He tells me he has checked with the secretary; the birds haven't been liberated yet. No point in my heading over there this early.

I feed the Ladies, walk Sally, our border collie, make breakfast. Back to the Internet. Now the forecast for Lyons is better: 46 degrees Fahrenheit and partly cloudy, winds out of the north at three miles per hour, gusting up to twenty. But the forecast for Southborough is light rain, westerly wind at two miles per hour gusting up to ten. And it might not stop raining till 5:00 p.m.

Mr. Lesieur calls me at 10:00 a.m. to tell me the birds were liberated at 9:00. But it could be slow going. Usually after a race of this length, the members of the club meet later that afternoon to compare the times on their clocks and calculate a winner. But the club secretary has decided on account of the bad weather that they shouldn't meet till 7:00 p.m.

The earliest we might see birds, says Mr. Lesieur, would be 1:30 p.m. "If we're lucky."

I'm leaving for Southborough before 11:00, and plan to get there by 1:00 to play it safe. But just three exits from Southborough, I'm stuck in traffic on the stupid interstate. I weave my little Subaru in and out of three lanes, desperate to escape the tangle. What if the birds come early?

I beg other drivers to let me ahead of them, and perhaps because it's Sunday, a churchly courtesy prevails: they give way. But there's only so much I can do. It's 1:15. I switch to an outside lane; 1:20: I dart into the exit lane; 1:27: I run one of two lights on Route 9; 1:32: I race along the twisty side road leading to the Lesieurs' street; 1:36: I park in the drive and leap from the car. Rather than knock on the door, I rush to the fence, find it locked, and peer through the slats.

There's a pigeon on the roof of the loft!

I run up the front steps and bang on the door. Through the window, I see the Lesieurs sitting in the living room with three guests. Mr. Lesieur motions for me to come in. Without greeting anyone, I blurt, "There's a pigeon here already!" Mr. Lesieur and I rush through the living room, through the dining room, through the sliding door to the back porch, and out into the backyard.

The pigeon is still out there. "Would you look at that!" he says. "C'mon, guy!" he urges the bird. "Get in there!" We rush to check the clock. The pigeon has already been recorded by the clock's electric eye: he's Number 149, and the clock reads 13:41.02.

But as it turns out, he's not the first bird. The clock shows another one arrived, unseen, at 13:22:03: Number 955, just a yearling.

"That was FAST!" Mr. Lesieur says. "That's four hours and twenty-two minutes! He wanted to get back to his hen!"

Mr. Lesieur pulls out the calculator to do the math: Number 149 flew from Lyons, through rain and winds sometimes gusting up to twenty miles per hour, at 1,895 yards per minute. He reminds me how fast that is: 1,760 yards per minute is a mile a minute. The liberator told me that from Southborough to Lyons is a six-and-a-half-hour drive.

But how does this first bird stack up against the other birds in the race? Have any of the others come in?

Mr. Lesieur gets on the phone, while I meet the guests. A nephew, Mike, is here, with his fiancée, Beverly. With them is a friend who hasn't seen Mr. Lesieur in fifty years: Al Duffy. A short, stout guy with a moustache and black hair, Al tells me that when they were kids, they used to catch pigeons from under the bridges together. They would sell them to Chinese people for thirty-five cents each. "That was a lot then. You could get anything you wanted for five or six pigeons. Candy cost a penny, and a movie was ten cents."

Mr. Lesieur reports back to us: "They all got 'em at one thirty-two! But Ardi Farhadi, he's closer than me, and Bibbo's loft is too short . . ."

He's not sure who's won yet, but one thing's for certain: his bird's in the running. Since last week, he says, "I came up from the bottom to the top!"

Outside, the other birds are coming in: Number 3106, with its two blue bars on each wing. Number 48, a pinkish yearling. Numbers 106, 127, 320, 10226 . . .

Mr. Lesieur is on the phone again calling the other racers. "I didn't even know we had birds!" he says in amazement. "Who the hell would know they'd come back in that length of time?"

We won't know it till later that evening, but Mr. Lesieur's pigeon came in second in the club. One of Dave Gage's pigeons beat him by ten yards a mile. Mr. Lesieur's Number 955 finished twenty-second in the Concourse, out of a field of 684 birds returning to sixty-one different lofts. A fine showing.

But he doesn't know that yet. Whether he's won or not, he says, "don't make any difference—you don't have to be first to be in pigeons. It's just the fun of flying."

At the moment, we face a long wait till the pigeon club meets at seven. It will be a festive evening, and I'm more than welcome there, now that I know the guys. But as Mr. Lesieur knows, my husband is flying to Italy tomorrow, a business trip of ten days. "Go home to your husband," Mr. Lesieur tells me.

I get pulled over for speeding on the way home.

Parrots

Birds Can Talk

O n my fifty-first birthday, I went dancing with a parrot.
  I have "danced" with quite a number of birds and beasts before. Playing music we both enjoy, I've spun around the floor with a cockatiel, lovebird, or parakeet perched on my shoulder. Holding (at different times) lizards, ferrets, and baby chickens, I've waltzed, Twisted, and Ponied to everything from Handel's *Messiah* to Springsteen's "Born to Run." But this would be different. This time, I would be dancing with Snowball—a twelve-year-old medium sulfur-crested cockatoo with an international reputation for his dance moves.

I had first seen Snowball on a YouTube video. Almost from the moment it was posted in September of 2007, the little movie of the dancing cockatoo went viral. More than two hundred thousand people viewed it in one week. And over the course of months, millions around the world thrilled to the sight of a fifteen-inch snowy white parrot with a tall yellow crest rocking out to his favorite song, "Everybody" by the Backstreet Boys.

"*Everybo-dy!*" exhort the lyrics. "*Rock your bo-dy!*" On the video, Snowball enthusiastically complies: from the back of a grey upholstered swivel chair, the cockatoo bobs his head, erects his sulfur crest, and jauntily waves one dark grey, clawed, prehensile foot in the air, then the other, all remarkably in time to the music. "*Rock your body, right!*" He varies his dance steps. He favors his left foot, as do most parrots, and while he usually lifts his right foot only once before switching to the other, he may wave his left foot once, twice, thrice, or more, depending on his artistic whim. Sometimes he sidesteps from one end of the chair to the other. Sometimes he faces the camera, and at others, as you might do with a dance partner, he turns his back. Sometimes he screams with delight—on a different note, but perhaps in the same spirit as some of the boy band's teenage fans. Snowball is obviously having a blast.

So are the folks who watch him. Snowball has danced for delighted audiences on *The Tonight Show*, the *Late Show, Good Morning America, The Ellen DeGeneres Show, Inside Edition*, CNN, FOX's *The Morning Show with Mike and Juliet,* and Animal Planet's *Most Outrageous. Time* magazine named his YouTube debut as one of the top ten videos of the year. He was featured on a Japanese quiz show called *Believe or Not* (in which two of the five panelists didn't believe a cockatoo could really dance like that). Lately, he stars in a Swedish television commercial for bottled water and has just filmed another for Geico.

"He *loves* to dance," his owner, Irena Schulz, told me when I phoned her to arrange a visit. "He's like the Energizer Bunny. He keeps going and going and going!" But this is not what makes Snowball unusual. So many birds bob, hop, twirl their heads, and flap to their favorite tunes that BirdChannel.com held a dance-off in which owners sent in videos of dancing birds so viewers could vote on those they liked best. (The winner: Poirot, an African grey, a silver-grey bird with a bald white face, yellow eyes, and crimson tail feathers who belongs to Mother Barbara of the Convent of St. Elizabeth in Etna, California. He bobs his head and waves a small wizard's wand to part of Vivaldi's *Four Seasons*.) What distinguishes Snowball is the way he keeps time with the music. Snowball's sense of rhythm is so refined it has garnered the attention of researcher Aniruddh Patel of the Neurosciences Institute in San Diego.

Perceiving rhythm is an act so natural to us humans that we tap our feet to music without realizing it. But it is actually a sophisticated cognitive feat, similar in many ways to skills we use in language. After all, as Patel points out in his book *Music, Language, and the Brain,* music and language share many similarities. Both are strings of organized sound, full of meaning for both performer and audience. Both powerfully manipulate emotions. Both have syntax—principles governing the combination of their structural elements into sequences. Both activate multiple areas of the brain. Both have richly structured rhythmic patterns, though the "beats" of speech don't mark out a regular pulse as most music does.

"Only certain types of brains" can perceive, create, anticipate, and synch to the beat of musical patterns, Patel writes. His careful studies of Snowball's dancing show that the cockatoo is capable of something scientists call beat perception and synchronization. In order to bob and strut in time to the music, Snowball must produce a mental model of the song's recurring time intervals and coordinate his movements with the model in his head—an ability never previously documented in any species but our own.

"Language and music define us as human," Patel writes in his book. Music, dance, and language are celebrated human universals. Even the Pirahá tribe of the Brazilian Amazon, he notes—whose language has no words for exact numbers, no fixed terms for colors, and who create no complex visual art—celebrate music and enjoy dance. Most of us would agree that music, dance, and language are all crucial to what it means, what it feels like to be human.

What does it mean that a bird can do something so fundamental to human experience and gain access to a source of such primal human joy?

In his lovely book *Birdsong: A Natural History,* Don Stap writes that "birdsong is like light streaming through the keyhole of a lost world." I feel the same allure in the chance to dance with Snowball. But perhaps it is not so much a lost world that I long to glimpse but a strange parallel universe—where creatures made of air and cloaked in feathers share with us the capacity for language, an appreciation for music, and even the ability to dance.

* * *

All of my own parrots appreciated music. My parakeet, Jerry, loved Herb Alpert and the Tijuana Brass and would chirp enthusiastically along with the songs, often while pecking with special vigor at his reflection in the hanging mirror in his cage. The parrots who came to me later in life—peach-faced lovebirds, two cockatiels, a crimson rosella—preferred female voices to brassy wind instruments, particularly Joni Mitchell. Her airy, high-pitched songs inspired the birds to sing their own tunes at the same time as hers. Only one of my birds imitated longer snatches of any song, and that was my yellow, white, and grey cockatiel, Kokopelli. But she didn't imitate the folk singer: she loved to whistle the National Geographic theme song and once helped me win a contract to develop and script a television documentary for the society. Fortuitously, Kokopelli, who liked to sit on my shoulder at my desk, whistled the song into the phone during a conference call to the director of the Natural History Unit of National Geographic Television. I think he took a chance on my proposal largely because he was so charmed by the bird.

Kokopelli was the parrot whom, other than Jerry, I loved the most. My two lovebirds enjoyed flying around my office, chewing my papers and posters, and would occasionally perch in my hair, but they were largely occupied with each other (at least till one killed the other and ate his head off—like people, even mated pairs fight). My other cockatiel, Octavio, and the rosella, Rozencranz, were elderly rescues. While I loved them, too, they were not overly affectionate. But I raised Kokopelli from a fledgling and we were very close. She and I showered together in the morning; she spent much of the day out of the cage as I worked and we would whistle back and forth to each other every few minutes. She made up a game we used to play with the little girls next door: one day Kate came into our living room carrying the long tail feather shed by one of our roosters, and Kokopelli flew to her from my shoulder. When Kate handed the feather over to her sister, Jane, to examine, Kokopelli flew to Jane. It became a game like capture the flag: whoever had the feather attracted Kokopelli. She would happily play her game with anyone who came in the house, even complete strangers, until she was panting from exertion and we decided to quit. I would then put the feather back in the crystal pen holder in the living room, signaling the end of the game. She never tried to fly to the feather in its holder, so it wasn't the feather per se that held her interest; clearly it was not a toy she wanted to play with by herself but part of a game that was only fun when played with others.

To my great sorrow, one morning, when she was only three, I found Kokopelli dead at the bottom of her cage. She had seemed in perfect health; such inexplicable deaths are heartbreakingly common in birds. But even in our short time together, she gave me a wonderful glimpse of the extraordinary playfulness and inventiveness of her kind.

For their intelligence, beauty, agility, and uncanny mimicry of human speech, parrots have been treasured as pets for millennia. Parrots are carved in pictographs on the walls of pharaohs' tombs. In the third century B.C., the philosopher Aristotle had a pet parrot. The species is unknown, but her name lives on: it was Psittace, from which scientists derive Psittaciformes, the scientific order to which parrots, which include the lovebirds, cockatiels, and cockatoos, belong.

Parrots are not closely related to any other bird group. They are an ancient lineage. (A recent fossil find of a cockatoo-sized parrot from 55 million years ago in Scandinavia predates most modern bird orders by more than 20 million years.) Parrots are extraordinarily distinctive: there's no mistaking a parrot for any other kind of bird. Some 353 species of parrots grace our planet today, from the three-and-a-half-inch buff-faced pygmy parrot of New Guinea to the eagle-sized hyacinth macaw. All of them have large heads, short necks, stout, strong, hooked bills, and muscular, agile tongues. Their toes are opposable, two in front, two in back, unlike most other birds except pigeons and woodpeckers, giving them a humanlike dexterity. Parrots are in general very chatty birds and excellent mimics.

Parrots' devotion to their mates is legendary. Less than 5 percent of mammals form any but short-term pair bonds, but fully 90 percent of all bird species during any breeding season are monogamous. Parrots take this to the extreme. In the wild, parrots mate for life, and because their lifespans rival those of humans, that can be a very long time. A parrot may form an extraordinarily close monogamous bond with his or her human, for as long as they both shall live. The seventeenth-century duchess of Richmond and Lennox so adored her pet African grey that she decreed that not even death should part them. After both died—the bird shortly after the Duchess in 1702—the parrot was stuffed and still sits beside the wax effigy of his mistress in Westminster Abbey. In the war of 1812, when the White House was set aflame, First Lady Dolley Madison could rescue only two things from the burning building; she chose one national treasure (sources debate whether it was the Declaration of Independence or a portrait of George Washington) and one personal treasure: her beloved green parrot, Uncle Willy.

Parrots reward their human partners with great loyalty. In *Of Parrots and People,* Mira Tweti tells the story of a parrot who died defending his owner, who was also killed in a robbery attempt. When the murderer was arrested, "he told the police he had never seen anything fight as hard as that parrot to save his owner." The killer was charged as a result of the human DNA on the bird's body, which had been chiseled by the brave bird's strong beak out of the offender's face and hands.

So it is no wonder that humans who live with parrots may love them deeply. But a relationship with a parrot is different, of course, from one with a spouse or a child, a dog or a horse, for although parrots may share our homes and own our hearts, they are not fellow mammals. They are birds, and this makes them wondrous and mysterious.

Yet of all the creatures on the planet, parrots alone might be able to tell us, in plain English, what it feels like to be a bird—and whether it is anything like being human.

In Schererville, a suburb in northwest Indiana, about an hour's drive from Chicago, I visit with Irena Schulz, her husband, Chuck, and their five-year-old son, Danny, who live with a changing population of anywhere from twelve to thirty parrots. Irena, a molecular biologist with luminous blue eyes and a spiky blond hairdo that recalls a cockatoo's crest, has turned the family's grey split-level home into a nonprofit sanctuary for unwanted parrots called Bird Lovers Only. On a bright Saturday morning in February, she opens its door to me and my companion, Field Museum scientist Gary Galbreath, and welcomes us inside.

Mookie, a large white Moluccan cockatoo with a feather-picking problem, issues an ongoing "BAAAAA" from her spacious, toy-filled cage. From upstairs, June, an abused macaw rescued along with two others, yells, "Cut the crap! Cut the crap!" When we enter Mookie's room, which she shares with seven other parrots, she ceases her sheep imitation and instead says cheerily, "Hello, Mookie!" Meanwhile, Ralph, a blue and gold macaw, holds out his wings like a host ushering a guest inside and says, "C'mon in!"

Snowball, the rock 'n' roll parrot, also talks, but for now, he says nothing. That's fine. After all, we have just met. I want us to be friends, but you can't rush these things. Before interacting with him, I expect to spend some time just sitting in his presence being harmless, talking with the humans. I hadn't counted on so much conversation from the other birds.

Snowball is downstairs in the den, inside a huge cage full of brightly colored wood and plastic parrot toys, alongside three other parrots in similar cages. The setup, as well as the timing of treats, television, and out-of-cage play on one of the many PVC parrot jungle gyms, changes from day to day. Irena explains that this wards off boredom, which can lead to feather picking.

Irena introduces us to Snowball's current immediate neighbors, three African greys. Two of them were owned by an elderly couple. Bandit, who has only one eye, speaks in the voice of the wife. "*Lar-ry!* Telephone!" he'll call. He's answered by Ben, whose legs are deformed, so he sits on a platform rather than a perch. Ben speaks in the voice of the beleaguered husband: "What? Yeah, O-K." Sometimes Ben imitates Larry in another way: he blows a raspberry. To which Bandit replies, in the wife's aghast voice, "Oh, my WORD!" The human couple had grown too frail to properly care for their pets, Irena explains, and since giving up the birds, the wife has died. But even after her death, the marital conversation continues in the voices of their parrots.

Mitzi is in the cage next to Snowball's. She, too, belonged to a couple, but bonded first with the wife. Mitzi speaks in her voice. Mitzi reproduces entire telephone conversations, complete with a perfect imitation of the ringing of the phone. "Brrring! Hello," she answers in the woman's voice. She proceeds to chat with a nonexistent caller, occasionally laughing and asking "Oh, really?" until the "call" starts to wind down. "Okay," says Mitzi, "see you Friday!" And then she imitates the click of a phone hanging up.

Talking birds have always enchanted humans. According to Pliny's *Natural History*, a raven who hailed the emperor Tiberius every morning was so revered that at the bird's death, he was honored with a funeral procession through the streets of Rome. The emperor Caesar Augustus had a parakeet who greeted him daily, and after his victory over Mark Antony in Egypt in 29 B.C., he purchased a raven whose trainer had taught him to say "Ave, Caesar Victor Imperator." (The trainer had wisely taught another bird to say "Ave, Victor Imperator Antoni" in case the battle went the other way.) Queen Victoria had an African grey who delighted her each morning by singing "God Save the Queen." During World War II, Sir Winston Churchill kept company with a blue and gold macaw named Charlie—or so says the man who claims to be Charlie's subsequent owner. Churchill's family disputes the claim, but when interviewed in 2004, Charlie's current owner insisted the bird, at 105, still recited expletives about Hitler and the Nazis learned from the prime minister.

But why should birds speak human words? In the wild, quite a number of birds—including mockingbirds, starlings, ravens, and Australian lyrebirds, totally unrelated to parrots—imitate the calls of other birds, frogs, and insects. They even mimic the sounds of machinery. The late, great artist, author, and bird-watcher Roger Tory Peterson once told me about a mockingbird who lived by his mechanic's shop and imitated the car motors. More recently, a male European blackbird terrorized a neighborhood in Somerset, England, by precisely duplicating the wail of ambulance sirens, the shriek of car alarms, and the nagging call of cell phones that went unanswered for hours, a cacophony that started at 5:00 a.m. daily and played throughout the spring. In other cases, birds incorporate our music into their own. In his book *Why Birds Sing*, jazz musician David Rothenberg reports that in the 1930s, an Australian flute-playing farmer in Dorrigo, New South Wales, kept a lyrebird as a pet, who liked to sing a fragment of one of the songs the man played. After a time, the man released his pet. For thirty years thereafter, lyrebirds in the adjacent national park incorporated elements into their songs that were heard nowhere else among their kind: flute-like notes that recalled two popular tunes from the 1930s. Another lyrebird, filmed by David Attenborough's documentary film crew, incorporated into his song the sounds of chain saws that were mowing down its forest. Unwittingly, the bird was, Attenborough said, "singing its own doom."

Ornithologists theorize that mimicry may have evolved initially to fool other birds or predators. A parrot who could mimic the sound of another bird whose nest it wished to usurp could entice the bird to respond, giving away its location. A bird who could mimic a predator might scare it off by suggesting that the territory was already occupied by another predator of the same species. As some of the earliest orders of birds, parrots may have invented mimicry. But other birds use the skill to great advantage, too. In songbirds like mockingbirds, mimicry allows a male to expand his song repertoire, a crucial advantage when females prefer the male with the most songs. Migratory species like marsh warblers broadcast the calls of African birds, possibly to inform potential mates where they spend the winter; it might be wise to choose a mate adapted to winter where you do, so your offspring will inherit these same adaptations. Thick-billed euphonias, tropical tanagers, use mimicry to enlist the help of other birds in protecting their nests. Imitating the mobbing calls of neighboring species enlists the neighbors' help to scare away predators that threaten their eggs or young.

In captivity, of course, human conversation, music, appliances, and broadcasts are all fodder for a mimic's repertoire. It's clear that many birds choose to imitate human speech simply because they like the sound of what they hear. After listening to a TV commercial, a captive starling was so taken by the phrase "Does Hammacher Schlemmer have a toll-free number?" that he began saying it just one day after hearing it for the first time. I can relate to that: in my limited phrasebook Italian, the sentence I love most to say—which sounds terribly momentous, if you don't know Italian, and features a bold, rolling *r* and a lovely string of alliterations—unfortunately translates to "I have broken my eyeglasses." Many people love to belt out the lyrics to songs they don't understand, sometimes even in languages they don't speak.

But thousands of owners say their birds also choose to do something completely different with the words they learn. Parrots' utterances are often delivered with uncanny timing in appropriate contexts. After a lost African grey named Yosuke was discovered perching on a neighbor's roof in the city of Nagareyama, near Tokyo, the parrot announced to his rescuers (in Japanese, of course), "I'm Mr. Yosuke Nakamura," and provided his full home address.

Many pet parrots greet their people with a cheery "Good morning!" and respond to a ringing phone with "Hello." Others like to call the family dog, using the voice of the owner. (When the dog obediently appears, the parrot lets loose peals of human laughter.) In some cases, parrots seem to use language to express profound emotion. The first time my friend Liz Thomas was bitten by her normally sweet yellow-naped Amazon, Viva, she was horrified. "Don't you bite me!" Liz scolded, and with uncharacteristic vehemence slapped the bird's cage with a newspaper. In a stricken, quavering voice, Viva cried, "Oh, my God!"

In St. Louis lives a man named Jim who has bipolar disorder with psychotic tendencies, and whose African grey, Sadie, keeps his violent outbursts in check. Neglected by a previous owner, Sadie had picked out her feathers, and Jim nursed her back to health. She has done the same for him. When he used to get upset, he would pace, hold his head, and tell himself aloud to calm down. One day Sadie started doing it, too. Now whenever he gets upset, the parrot soothes and calms him, saying, "It's okay, Jim. Calm down, Jim. You're all right, Jim. I'm here, Jim." He has never had a violent outburst in her presence.

Sadie and Jim may be extraordinary, but parrots who speak meaningfully are, in fact, remarkably common. On my flight from New Hampshire to visit Snowball, I happened to sit next to a man who told me about a cockatoo he knew named Mickey, also from St. Louis. Mickey was a smart bird who often opened and escaped from his own cage while his owner was out. One day the owner came home to find, to her great alarm, her Labrador retriever holding Mickey in his mouth. "Drop the bird!" the woman screamed. "Put Mickey down!" From within the dog's jaws, the bird cried, "Put Mickey down!" The dog, astonished, dropped the parrot at the owner's feet.

Snowball tends to express himself with dance more than with words, Irena tells me. This is merciful; at one point the parrot cacophony in the house grows so loud that Gary flees briefly to a bathroom to escape. But as we sit sipping tea on the couch in the cozy den, surrounded by the exclamations and murmurings of her parrots, Irena relates a story about something Snowball said recently that seemed to reveal a great deal about his feelings at the time.

Just a few weeks before my visit, an Animal Planet camera crew was filming Snowball at a dance studio in Chicago. Snowball was "instructing" the children in the dance class; they were supposed to copy his every move. As with all his public performances, Snowball seemed to relish the attention. "I want to say he's a diva," Irena confesses. "I know this is really strange, but that bird *knows* when he's on camera. If you're paying attention to him, he's fine and dandy. He poses for photographs. He loves being center stage. But if you divert your attention from him, he gets angry." Snowball's anger is easy to read. His movements become jagged. He bites and screams. Once the cameras turned to the children, Snowball began screaming. Usually he screams for attention in a high-pitched, insistent call that sounds vaguely mechanical, sort of like a cross between a smoke detector and a car alarm. But this time, Snowball was yelling a phrase he had probably heard from his previous owner's adolescent daughter: "It's not *fair*! IT'S NOT FAIR!"

Did Snowball know what he was saying? Why did he choose to use *these* words at that moment and not any of the many hundreds of others—including words that describe very different kinds of displeasure—that he has surely heard? What do our words mean to the birds who use them? How much of our language do they understand? How much of our culture—language, music, dance—might they mirror in their own?

So far, the one who has come closest to giving us the answers is an African grey named Alex.

Alex was an ordinary pet store parrot when Irene Pepperberg, a dark-haired beauty with a newly minted Ph.D. in chemistry, a photographic memory, and a passion for birds, bought him at a shop near O'Hare Airport on June 15, 1977. By the time he died unexpectedly on September 6, 2007, at age thirty-one, he was one of the most celebrated personalities in the world—a bird who changed forever our understanding of how nonhuman animals think.

To this day, an online community of Alex's friends and followers hold candlelight vigils on the sixth of each month in his memory. His obituary made headlines from the *New York Times* to the journal *Nature* to the *Hindustan Times*. People sent condolences from around the world. When I heard the news, I wept and then wrote a memorial commentary for public radio's environmental show *Living on Earth*. Though I had never met him, I had been following Alex's career for more than a quarter century.

Alex's name is an acronym for Avian Learning Experiment. Irene began her work with him at an exceptionally inauspicious time. She had earned her doctorate as a theoretical chemist, but a series of television documentaries on animal communication changed the course of her career. Unfortunately she entered the field just as trashing previous animal language experiments became the scientific vogue.

When Pepperberg began working with Alex, the best known experiments had focused on chimps, bonobos, gorillas, and an orangutan, our closest relatives. Although apes make a variety of specific calls, because the position of their larynx is too high in the throat, they cannot easily pronounce words (one chimp, after years of training, finally managed "mama" and "cup"). Researchers instead taught nonhuman primates sign language, or to use plastic chips or lexigrams on computer keyboards. I have always followed these experiments avidly, always cheering for the animals as they use these nonspoken languages to ask and answer questions, refer to objects present and absent, invent new words, and reveal their emotions. With the same gestural language that deaf Americans use to communicate, Washoe the chimp labeled a piquant radish she tasted for the first time as "hurt cry food." Koko the gorilla mourned the loss of her pet kitten, signing, "Cry, sad, frown." Many scientists, however, refuse to believe that animals have any sort of consciousness; some even deny that animals feel emotions—or suffer from pain—in a rigid adherence to Cartesian prejudices about human superiority to every other creature on earth. So, not surprisingly, critics claimed that the animals in the language experiments had not been tested rigorously enough and that the trainers had inadvertently cued them. At a widely publicized 1980 conference at the New York Academy of Sciences, the organizers denounced the signing apes as circus performers and even hired a magician to perform to denigrate the experiments. They claimed that nothing these animals did showed them capable of complex thinking.

Since most scientists doubted that nonhuman primates could use even the rudiments of language, they dismissed as ludicrous the idea that a bird with a brain the size of a shelled walnut might prove what apes could not. Yet this was Irene's idea, and to me, it made perfect sense. After all, parrots can talk.

Birds don't create sound in the same way we do. Yet parrots—African greys in particular—can speak human words with astonishing clarity. Parrots and humans use different physical structures to create the same sounds. Humans speak and sing by vibrating the membranes of the larynx, our voice box, on exhalation, and sometimes on inhalation; we modify the sound by movements of the lips, tongue, sinuses, and inside of the mouth. Birds lack lips, and their larynx doesn't make any sound at all. The bird's voice box is the syrinx, a cartilaginous structure located at the juncture of the two windpipes, or bronchi, that houses paired, vibrating membranes that produce sounds. Songbirds can use the right and left halves of the syrinx independently and produce two simultaneous sounds, as if singing harmony with themselves—which you can hear especially well in the spiraling, liquid song of the hermit thrush. Parrots can't do this, but they have other adaptations that make them particularly good talkers. For instance, a parrot's windpipe is slightly flexible, while the human's is stationary—which is why humans speak the sound for *a* with an open mouth, but a parrot says it with an almost completely closed beak. A parrot can move the tongue, beak, glottis, larynx, and esophagus to create the sounds we make with our lips, as for the letters *p, b,* and *d.* And because of the combination of delicacy and force necessary to pry seeds, a parrot's tongue (which is not an organ for tasting; a bird's relatively few taste buds are mostly on the palate and lining of the mouth) is exceptionally flexible and strong.

These natural abilities would seem to make parrots ideal subjects for language experiments, but Irene's initial grant proposals were rejected—often in language that implied she was insane. In the 1980s, even those mavericks who hoped interspecies communication was possible felt it might work only with fellow primates, or at least large-brained mammals like dolphins. Birds were out of the question, because their brains are far smaller and fundamentally different from ours. In mammals, the thick, deeply wrinkled cerebral cortex is the central feature of the foremost part of the brain and the seat of human thought. In birds, what was once thought to be the cerebral cortex is thin and superficial. Researchers discovered they could surgically remove that area of a bird's brain with hardly any effect on behavior at all. Most scientists assumed from this result that birds weren't very bright.

The center of higher learning for a bird, it turns out, is a part of its brain that doesn't look cortical at all but is nonetheless derived from the same precortical areas as those of mammals. It was once called the hyperstriatum. Irene likens a bird's brain to a PC and a human's to a Mac: even with different wiring, bird and human brains can do many of the same things. Despite the fact that many pet parrots seemed to speak meaningfully, attempts to teach birds meaningful use of language in the laboratory failed, because the standard laboratory procedure was to starve the birds to 80 percent of their normal weight (ostensibly to motivate them to work for food rewards) and to isolate them in sterile cages so that scientists controlled all stimuli (supposedly to focus their attention). These methods were not only cruel but staggeringly stupid. "Isn't it blindingly obvious," Irene asks the readers in her wonderful memoir, *Alex and Me,* "that communication is a social process, and that learning to communicate is a social process too?"

Irene taught Alex according to a completely new paradigm: he was not starved or isolated; he learned in a rich, stimulating, loving, social environment, as young humans—and more relevant, as young birds—do. For, although songs are instinctual for some species of birds, including doves, cuckoos, and North American flycatchers, this is not the case for parrots, hummingbirds, lyrebirds, and the largest bird order of all—the passerines, or songbirds, who constitute the most diverse taxon of birds on the planet, with 5,400 species. They *learn* their songs, and they do so in much the same way children learn language: They listen carefully to adults. They practice what they hear—not just the order of the notes, but the order of the songs, if they have a repertoire. They may recombine elements in new ways, and the learning process may take a long time: some species of lyrebirds take up to six years to learn a song.

Alex usually worked with two trainers. One trainer would ask a question: What toy? And the other would give the answer: Truck! Alex watched as correct answers were rewarded and incorrect ones were not. Importantly, when Alex spoke a new word Irene was teaching, he was not given food in response; he was given the object he named, so the word and the object would be linked in his mind. As his vocabulary expanded, Alex would request the reward of his choice. "What do you want?" the trainer would ask, and Alex learned to reply according to his whim: "Want walnut." "Want corn." "Want shower." "Want grain." "Want banana." "Want water." "Wanna go shoulder." (Sometimes he combined elements of the words "pasta" and "treat" with a result that might have upset campus security: "Want pot!") He clearly understood what he was asking. If Alex was presented with a reward other than the one he wanted, he would say "Nuh!"—and sometimes throw the mistaken item to the ground in disgust before repeating his request.

Alex learned that speech was power. He did not wait until he was asked to announce his desires. If he grew bored with a lesson or experiment, he would say so, announcing "Wanna go back" (to his cage) or telling the researchers to "go away." When Alex wanted a nut, he expected a nut—and showed his irritation when the service was slow. During one test he found particularly boring, Alex announced repeatedly "Wanna nut," yet the questions continued. "Wanna nut," Alex said again, to no avail. In a previous experiment, Alex had been learning phonics, so finally he figured he had better spell it out. "Wanna nut," he said again, and then pronounced, clearly and slowly, "Nnn-uu-tuh."

Irene stresses that her objective with Alex was not to teach Alex human language per se. It was important that he understood what he was saying, but she did not concentrate on teaching him perfect grammar, tenses, or the finer points of syntax. Human speech was merely a tool to help her toward a larger goal: to probe the mind of a bird, to reveal concepts he understood, to show us something of how he saw the world.

Her efforts succeeded to an extent that surprised even Irene herself. Presented with a tray containing a green spool, a green truck, and a green pom-pom, if asked "What's the same?" about all the objects, Alex could tell you: "Color." Asked *what* color, he would say, "Green." And what is the green pom-pom made of? "Wool." Yet these were the least of his abilities.

Alex asked questions about the world around him: "What's that?" he queried a student when he saw his reflection for the first time in a mirror. "That's you," she answered. "You're a parrot." Alex then asked what color—and learned the word "grey." Too, like some of the language-trained apes before him, Alex invented words: "banerry" for the cherry-colored, banana-flavored fruit we call an apple, "corknut" for an almond in its porous-looking shell, "yummy bread" for cake, "rock corn" to differentiate hard, dried corn kernels from sweet, fresh corn. He could also count up to eight—a higher number than Liz Thomas found Kalahari Bushmen used when she lived among them in the 1950s. Alex recognized and understood Arabic numerals and he spontaneously figured out how to add.

Alex wondered aloud about what others thought. When he was hospitalized several nights at the vet's for an infection, Alex's cage sat next to the accountant's desk. One night when the woman stayed late working on the books, Alex asked her, "Want a nut?" "No, Alex," she replied, and went back to working. "Want corn?" he asked her. "No, thank you, Alex, I don't want corn." She continued to ignore him. But he persisted. Exasperated, and in a petulant voice, he asked, "Well, what *do* you want?"

Occasionally, Alex used speech to tell Irene what was deep in his birdy heart. One of the most telling incidents took place in Arizona, where Irene and Alex worked in the 1990s. One night she brought Alex home from the lab to enjoy a change of scenery. Suddenly Alex began to scream: "Wanna go back! Wanna go back!" Through the window, he had seen a pair of screech owls building a nest in the roof over the patio. He kept screaming even after she closed the drapes so he could not see them; he knew the owls were still there and continued to beg her to take him away from them, back to the safety of the lab.

Alex had never seen an owl before. Yet he knew, somewhere in the depths of his ancient genetic memory, that these birds were predators of parrots— and was communicating this prehistoric knowledge to a late-twentieth-century scientist, in spoken English.

Irene's work with Alex "has revolutionized the way we think of bird brains," said Diana Reiss, a psychologist at Hunter College who studies dolphins and elephants. The term, she said, "used to be a pejorative, but now we look at those brains—at least Alex's—with some awe." In the years since Alex left the pet store for the lab, a major revolution has taken place in the understanding of the minds of birds. At the 2009 meeting of the American Association for the Advancement of Science, University of Cambridge professor Nicola Clayton pronounced the pejorative "birdbrain" obsolete; instead, she told her audience, we should talk about "brainy birds."

There are still skeptics. Clive Wynne, an animal psychologist at University of Florida, recently told a *New Yorker* reporter profiling Irene and Alex's work, "If there's really something going on there"—and here he stresses that Alex was studied for thirty years—"then somebody else, somewhere along the line, would have replicated this. If we really want this to be science and not just some sort of adjunct to the entertainment industry, we shouldn't be relying on one animal." Why, he asks, is Alex "the only parrot in the history of parrotdom" to show cognitive powers previously accorded only to humans?

The Backstreet Boys seem to ask the same thing.

"Gotta question for you," go the lyrics to Snowball's favorite song. *"Better answer now."*

"Ye-ah!" agrees the chorus.

"Am I original?" the lyrics ask.

"Ye-ah!" the chorus answers. When you watch the video of Snowball rocking out to "Everybody," it's tempting to think the song was written about him. Snowball, crest erect, facing the camera and bobbing enthusiastically to the beat, seems to think so.

"Am I the only one?"

"Ye-ah!" sings the chorus. But Snowball's dancing is providing science with a much broader answer.

Snowball came with a CD.

His previous owner gave him up, Irena tells me. Snowball had begun to attack the man's daughter. Perching peacefully in his cage in the warm, sunny den, Snowball does not look like an attack parrot; by now, he is so relaxed that he is slowly running his left claw through the yellow feathers of his crest like a comb, the picture of contentment. But Snowball had a rap sheet. He'd fallen in love with the man's daughter, Irena explains, as parrots often do at sexual maturity. Alas, Snowball's chosen mate had committed the unforgivable sin of moving away to college. No decent parrot would ever think of leaving a mate for days, much less weeks, so of course Snowball was incensed. Every time the girl returned, he flew at her and tried to bite her. Because a cockatoo's beak is built for opening nuts that humans need hammers to crack and is capable of exerting four hundred pounds of pressure per square inch, Snowball's anger, no matter how justified, became a serious hazard. So Snowball's owner reluctantly relinquished him to Bird Lovers Only. When he did, he also handed Irena a CD. He didn't say much about it.

"He told me to play it and watch Snowball's reaction," Irena tells me.

"When I did, I nearly fainted." Snowball's favorite song turned him on like a light switch: immediately he began dancing, with a gusto and precision that took her breath away. "I thought, What on earth is going on? I'd never seen anything like it."

Neither had Aniruddh Patel.

A senior fellow at the Neurosciences Institute, he's president of the Society for Music Perception and Cognition, a nonprofit group of researchers in psychology, music, and cognitive neuroscience. He had long been fascinated by what seemed to him, from an evolutionary standpoint, the "peculiar phenomenon" that language and music "appear in only one species: *Homo sapiens*," as he wrote in his 2008 book.

In his extensive research, the scientist had scoured the literature and plumbed his numerous contacts for reports of nonhuman animals spontaneously moving to the beat of music. He had even gone to Lampang, Thailand, to visit the Thai Elephant Orchestra, where he witnessed elephants who had been taught to toot harmonicas, play giant xylophones, and hold wooden mallets in their trunks to strike drums and gongs at a remarkably steady beat. But, importantly, when performing in their ensemble setting, the elephants showed no evidence of synchronizing their drumming to a *common* beat, which demands a cognitive talent of a higher order. Likewise, the rhythmic chorusing of katydids, for example, doesn't qualify as evidence of beat perception and synchronization, he argues; these animals lack the flexibility to synchronize to rhythms at a wide range of tempi, as humans can easily do. Actually, Patel points out, in some cases the insects are not trying to sing in synchrony at all; they just end up sounding that way because all the males are attempting to call *first*.

But then, from an assistant working with neurologist-author Oliver Sacks, one day the California researcher got an e-mail with a link to Snowball's video. (Sacks references Snowball in a footnote to the updated paperback edition of his 2007 *Musicophilia: Tales of Music and the Brain*.) Patel was intrigued—but skeptical. Could a cockatoo really keep the beat like that? He contacted Bird Lovers Only and made arrangements to visit in April 2008, to see for himself.

By that time, Irena was used to getting calls out of the blue from strangers who wanted to meet Snowball. Martha Stewart's TV producer had called, wanting Snowball on the show, but only if Irena promised not to do any other show first. "But ten other shows had already called with the same request!" Irena says. The day after Snowball was on *The Morning Show with Mike and Juliet* in New York, *Inside Edition* called and asked if they could film Snowball. "I asked, 'Well, when were you thinking you would do this?' And they said, 'How about half an hour from now?'" Irena remembers. "I was like a deer in the headlights!"

Irena is grateful for all the publicity; Snowball makes a great spokesparrot for the other abandoned and relinquished birds, and the exposure has brought in much-needed funds for her sanctuary. But Ani Patel's call was even more exciting. Irena, a talented alto and guitarist who often performs for her Methodist church, had given up her career as a molecular biologist to care for her birds and her son. She tells me: "I love science. I love birds. I love music. And here, because of Snowball, all three of these have come to me!" Irena and her husband were coauthors with Patel and three other colleagues on the first scientific paper on Snowball, presented at the Tenth International Conference on Music Perception and Cognition in Sapporo, Japan. More papers are in press. The collaboration, she hopes, will continue for a long time, and Patel feels the same way—for the questions that the first study asked have yielded surprising and intriguing answers.

Was Snowball really synching to the beat of the music, or was he getting timing cues from Irena? Her dancing shadow appears on the wall in the video—and who can watch Snowball dance without dancing herself? Even more important was to see if Snowball would dance to more than one beat. Could he adjust to match others?

To find out, while Snowball danced, everyone suppressed his or her own head bobbing and foot tapping (with great difficulty). Using software, Patel and colleagues manipulated Snowball's favorite song to create eleven different versions at different tempi but the same pitch. Irena played them all to see what Snowball would do and videotaped the results.

I'll get to see what happened firsthand, Irena promises; she still has the altered music on the computer, and Snowball will happily give me a demo when we dance together later this afternoon. But I have already read the paper, of course, and know the outcome. The videos were carefully viewed in sixty-frame-per-second time resolution and scored by coders who watched with the sound off and didn't know which tempo was being tested. Their findings: Snowball really *does* dance to the beat, and he does it for sustained periods. He adjusts his dancing to altered tempi. He prefers the original tempo or its faster versions. He doesn't do as good a job slow-dancing to a rock song. (Neither do many people; that's why most rock songs are fast.)

The conclusion to the paper: "This study shows that beat perception and synchronization is not a uniquely human phenomenon." Why do we see beat perception and synchronization in a parrot and not a fellow primate? The findings strongly suggest that beat perception and synchronization are latent abilities "in certain types of brains . . . those that have been shaped by natural selection for vocal learning." The studies of Snowball strongly suggest that, despite our widely separated evolutionary history, the way our brains process information about sound may be even more like parrots than like our fellow primates.

That first study points to a series of others: What range of music does Snowball synch to? What influence do social cues have on his abilities? Is he better when dancing with a partner? Do Snowball's different movements (head bobbing, foot lifting, side-to-side head movements) mark out different levels of the metrical hierarchy? What is the relationship of his dance movements to the natural display movements of cockatoos?

Irena is already at work on this. "Snowball actually enjoys quite a number of songs," she tells me. Besides "Everybody," among his favorites are "Another One Bites the Dust" by Queen and "Edge of Seventeen" and "Stand Back" by Stevie Nicks (whose brother owns a cockatoo like Snowball, with whom she appears on the cover of her album *Bella Donna*). Luckily, these are songs that Irena enjoys, too. ("What if he adored rap music?" muses Gary in horror. "If it were me, I'd have to move out!") But you never know what might captivate a cockatoo: once, Snowball spontaneously started dancing to a TV commercial. Because of his fame, "people mail us different CDs for him to dance to," Irena explains. "One person sent in a CD of German polka tunes. I put it in the CD player and oddly enough, not only was he dancing to it, but two other birds around him started dancing!"

Since Snowball arrived in August of 2007, he has invented no fewer than twelve distinct dance moves. Each involves different movements, mostly of the head and feet. Sometimes he erects his crest and seems to do a head-banging dance like those popular at punk rock clubs. Sometimes he sways his body from side to side and lifts his crest simultaneously. Snowball's latest invention, Irena tells me, features a move in which he appears to be blowing kisses: he brings his claw to his beak, then turns both his head and claw in opposite directions, all in time to the music.

What is going on here? Where does Snowball get this talent? Why does Snowball dance? What does he experience while he is dancing?

Irena has given a great deal of thought to this. His previous owner, she explains, told her he had noticed Snowball bobbing his head to "Everybody." He and his children encouraged his movements and would dance with him, moving their arms around. Snowball soon added leg lifting to his bobbing. But this doesn't account fully for his unusual talent, much less his syncopation skills.

Irena suggests one possibility. "Maybe he was flying and hit his head." For their own safety, Irena trims her birds' wing feathers so they cannot fly and injure themselves. But before he came to her, Snowball was flighted. Injuries are more likely to result in disability than the emergence of a new talent. But strangely, this is not always the case. Irena and I had both read Oliver Sacks's *Musicophilia* and vividly recall two of his stories of patients whose brain damage triggered a new delight and hunger for music. One patient, a forty-two-year-old surgeon, was struck by lightning. Shortly thereafter, he developed such a craving for piano music that he learned to play. He started to hear music in his head and found it so compelling that he taught himself musical notation in order to write it down. When he performed in concert, he electrified his audience. Another patient, a research chemist who had previously had no special interest in music, found that after brain surgery for a tumor, music moved her to rapture or tears. Could Snowball be similarly afflicted—or blessed?

At times, Irena says, Snowball behaves as if he is having a seizure. Over in his cage, right now, Snowball is playful and happy, picking up a yellow, pink, and orange plastic chain toy in his beak and tossing it around, listening to its clink and watching its movement. But sometimes, seemingly out of the blue, he spins his head and squawks wildly, out of control. She tells me of an incident that happened just after the camera crew had finished filming the Swedish water commercial. "They had filmed Snowball for four hours, from every angle. He was posing for the cameraman's kids, who were taking photos. And after that . . . he went beyond head banging. Way beyond that. It was frightening. He wouldn't look at me. I couldn't get his attention. It was the most frightening thing I've ever seen—even though I've seen other seizures." Irena is not a stranger to neurological disease. Her specialty as a molecular biologist led her to researching cell injury and cell repair at a lab at Rush University Medical Center that worked with Christopher Reeve on paralysis and Michael J. Fox on Parkinson's disease. "I don't know what's wrong with Snowball," she says. "Is he epileptic? Is he Touretting?" (Or, suggests Patel, who has seen video of Snowball's fits, is he having a particularly dramatic tantrum? "Cockatoos are emotionally complicated. It's hard to know what's going on in their heads.")

Perhaps Snowball is a savant. The human savants whom Sacks describes in his book lack most normal mental skills but have remarkable faculties for calculation, drawing, or—most notably—music. One savant, a retarded man who could not retain the meaning of anything he read, could nonetheless flawlessly remember two thousand operas and replicate every note every voice sang or any instrument played on his piano. I'm not suggesting Snowball is retarded, but his normal creative life, like that of all caged birds, necessarily is. This Irena well knows. In the wild, parrots live rich, demanding, interesting lives. They may fly fifty miles a day. They may visit many dozens of fruiting trees, keeping in mind long-term memories of the fruiting schedule and location of each throughout the year. They may have social interactions with dozens or hundreds of different individuals, including birds and animals of species other than their own. "These birds are wild animals," she says. "They don't belong here. Even though I am here, playing with them, feeding them, cleaning them, I'm only giving them one percent of what they need." Even parrot trainers I have spoken with have privately admitted that parrots are so psychologically complex and idiosyncratic that few can live happily in human homes.

Possibly Snowball dances as an outlet for creative energies that otherwise would be expressed naturally in the wild; possibly he dances to relieve neurological symptoms. Either way, says Irena, "music is therapy for Snowball." And it's powerful therapy indeed. People with neurological problems often benefit from music therapy more than they do from drugs. Those afflicted with Parkinson's disease, for instance, often become frozen in their movements; music can release them. Athletes often use music to coordinate their movements; once when Sacks himself was badly injured in a climbing accident on a mountain in Norway, he imagined the "Song of the Volga Boatman" to "row" himself down the mountain before nightfall using his arms and one leg. "I wonder," Irena asks, "is it possible that Snowball uses music to alter his own brain? Is he using dancing as a way to ward off seizures or in some way control them?" If so, Snowball's story may be showing us that the healing power of music crosses species lines.

Hearing of Snowball's seizures concerned Patel for several reasons. Naturally, as a researcher who hoped to describe what might be a widespread phenomenon, "I was worried he was doing this because he had a messed-up brain," he says. But no; Patel is convinced this is not the case. A paper published back to back with Patel's in *Current Biology* documents beat perception and synchronization in no fewer than fourteen species of parrots—from peach-faced lovebirds to blue and gold macaws—based on video of other avian dancers, most of them on YouTube. "This suggests *a lot* of birds can do this," Patel tells me in a phone conversation. "Snowball is definitely special," he says, "but he is not the only one." Snowball is special because he is *not* unique.

For Patel, Snowball has been the keyhole through which he has glimpsed minds he didn't know existed. Working with Snowball, he says, "has really made me think about how other species experience music, how another species responds to music." Now, he realizes that parrots—and likely all birds—"are hearing the world in a very complicated way. They have a lot of flexibility in their auditory perception—similar to the way we do it. It's suggestive of complex thinking. Snowball has taught us that other species do respond to human music. It has really turned my head around."

Patel has been in touch with Pepperberg's colleague Adena Schachner, the lead author on the other *Current Biology* paper. On a Wednesday afternoon one September, Adena joined Irene and Alex in the Brandeis University laboratory that had become Alex's new home. Adena had just begun researching the origin of musical abilities; she wondered what kind of music Alex liked. She filmed him as she played some 1980s disco. As Alex bobbed his head in time with the music, Adena and Irene both danced along.

* * *

That was September 5, 2007. Alex died sometime the night of the following day. But though his cage sits empty in the corner of the tiny ten-by-twenty-five-foot Brandeis lab, the work he began more than thirty years ago continues.

"All right, Griffin," Neil Dean, a tall blond senior majoring in business, says to the thirteen-year-old African grey perched on the T-shaped stand in front of him. Griffin, named partly for Donald Griffin, a pioneer in animal cognition research, and partly because he looked like a gryphon as a chick, has been with the study since he was seven weeks old. "Work is fun, Griffy!" Neil encourages him. "Nice and easy, buddy. You've got nuts coming your way!"

"What color, Griffin?" Brandeis students ask the young African grey.
He's still learning his colors.

Neil sits on a tall green stool next to Jessica Nu, a senior in black pigtails majoring in biology and environmental studies. Irene is away most of this semester on her book tour; while she's gone, students carry on teaching and testing Griffin. Griffin sits in front of them, slightly below eye level, on his T stand. Between the students sits a plastic utility box full of objects. Neil pulls from it a small piece of newspaper and holds it beneath Griffin's beak. "What matter, Grif?"

"Paper," Griffin replies softly, in a voice that sounds singsongy yet timid.

"Good! And what do you want?" asks Neil.

"Nut," the parrot replies. Neil hands him a small piece of cashew. Next, Neil holds up a green pom-pom, the sort of little balls that used to adorn the backs of tennis socks. "What matter, Griffin?"

"Wooooooooooo-lllll!" he shouts, his voice like a child enjoying a ride on the Ferris wheel.

"Smart boy!" says Neil. "What do you want?"

"Nut," Griffin answers again.

"Grif," Neil says next, holding up a large metal screw, "what toy?"

"Nay-ul," Griffin says softly.

"What do you want?"

"Nut."

Neil shows Griffin a plastic *S,* the magnetic kind that parents of young children have up on the refrigerator. "Griffin, what sound?"

"Sssssss," he hisses. And requests a nut.

Griffin correctly identifies a key, a rock, a toy truck. He's shown a plastic *U.* What sound? "Ooooooooo," he says softly. He's rewarded each time with a piece of nut.

Now Neil holds a small plastic yellow cup up to Griffin's beak, touches it, then pulls it slightly away. "Grif," he says, "what color?"

"Green," the parrot mutters.

"No, that's wrong," Neil corrects him.

"Yel-low," says Griffin.

"Yes! What do you want?"

"Nut," replies the bird.

Now Neil holds up a red plastic cup. "Grif, what color?"

"Gr-EEN," replies Griffin. And before he can be corrected, he says, "OR-ange."

"No! What color, Griffin?" Neil asks again.

"Blue."

Frustrated, Neil picks up the blue cup and holds it in front of the parrot. "No, *this* is blue." He puts it down and holds up the red cup again. "Try, Griffin, try. What color, buddy?"

Griffin is silent.

"When he gets something wrong, we model it for him," Neil explains to me. Jessica now steps in and answers the question for Neil. *"Rose,"* she says slowly and emphatically. (Alex was taught to say "rose" instead of "red" because "red" sounded too much like an earlier word he had learned, "peg." Griffin keeps up the tradition.) *"Rose,"* Jessica repeats.

"Good parrot!" Neil tells Jessica. "What do you want?"

"Want nut!" Jessica replies, and when Neil hands her one, she pretends to devour it greedily, smacking her lips. "Yummy nut," she says, "for *rose!*"

The students would still like Griffin to say "rose."

"C'mon, Griffin—try!" Jessica begs. She puts three pieces of cashew inside the red cup and shows them to him. "What color? What color?"

Griffin says nothing. He looks at us sideways with one sleepy, blinking yellow eye and stands on one foot, as if he's about to tuck his bill into his wing. He looks like he is thoroughly bored.

"C'mon, Griffin—try!" Arlene Levin, the lab manager, shouts from the front of the room. Arlene, a powerhouse of energy with fluffy auburn hair, brown eyes, and a tattoo I can't quite make out on her shoulder, but which I suspect might be a parrot, has managed the lab since Irene and her parrots moved to Brandeis from MIT six years ago. "They're like three-year-olds," Arlene tells me. She and her husband, Dave, have six parrots of their own at home. "It can be a game to him to get it wrong."

Griffin mutters, "Gr-EEN."

"No, not right," Neil says firmly. "C'mon, Griffin, you know this!"

In fact, he does: in a later session, one of half a dozen or so I will attend at the lab, Griffin is shown a series of two different colored cups on a platter and told which color to choose by touching it with his beak. He scores 85 percent correct—the equivalent of a B plus—and gets "rose" right every time.

Neil and Jessica give up on the red cup. They try a blue one. "What color, Griffin?" they ask with unflagging enthusiasm. "What color?"

"Gr-EEN! YEL-low! OR-ange!" Griffin cries in quick succession. He's just naming random colors—all except the right one. Except, of course, his words aren't really chosen at random. "You notice he always says the name of a color," Arlene points out. "He understands the category."

Griffin has been working on these sorts of questions for six years. He knows his colors. He knows shapes—two, three, four, five, and six corners. He knows the names of a number of toys and can identify matter ranging from paper to rock. He is learning phonics. He can make comparisons of a sort that young children commonly fail: he can tell you which object is smaller and which is bigger—even when an object that is bigger in one trial is the smaller object in the next. But Griffin will not tell us the color of the blue cup today. Why not?

Alex, as it turned out, often did the same thing. Orange was his particular bane; he seemed to confuse the color with both yellow and red. His mistakes were instructive. Irene figured out that parrots actually see colors differently than we do. Their visual palette is far richer and more complex than ours. They see more colors than we can even know exist, and scientists believe that to birds, colors are so vibrant that they seem to sparkle and shimmer. Many if not all birds see in the ultraviolet spectrum, and parrots probably do, too. One day Irene painted all Alex's orange objects from the same paint can, and the problem disappeared.

But in other cases, Alex's wrong answers had another cause. Irene could never prove it, but she is certain that he sometimes purposefully gave wrong answers—the avian equivalent of a practical joke. Irene remembers one day she was showing off Alex's cognitive prowess to an out-of-town visitor. She showed Alex a tray of objects of different materials and different colors. "How many green wool?" she asked. Alex eyed the tray, which among its contents contained only two green wool pom-poms, and, giving Irene his "wry" look, declared, "One."

"No, Alex," she said, exasperated. This wasn't a hard test for him. She knew he knew perfectly well there were two green pom-poms. "How many green wool?" And this time, he replied, in his two-syllable, sing-songy way, "Foo-wah."

And that's the way he continued to answer the question: "One." "Foo-wah." "One." "Foo-wah."

"Okay, Alex," Irene told him, "you're just going to have to take a time-out." She took him to another room and shut the door. As she recalls in her memoir, immediately from behind the closed door came Alex's plead-ing voice: "Two . . . two . . . two . . . I'm sorry . . . come here! Two . . . come here . . . two!"

"That was one of the things that surprised me most about working with Alex," Irene would later tell me, "purposefully giving wrong answers. I didn't expect the degree to which Alex would mess with my head!" But folks in the lab suspect that Griffin's mistakes spring from a different source. While almost certainly intentional, they may not reflect mischief or boredom. They may be a result of grief.

You can see something's been bothering him. In June, he plucked the feathers from his chest and belly and he still looks bedraggled as they grow back. He misses Irene, who raised him from a chick. Lately, she's been gone a lot. Ever since her last book came out, her schedule has been a whirlwind. These days, Griffin sees Irene mostly on Arlene's computer screen. When she plays a video of one of Irene's television interviews, Griffin cranes his neck to see. "Being called a bird brain is actually a compliment," Irene is telling the show's host. "It's a big breakthrough." Griffin recognizes her voice. Arlene moves his T stand so he can get a better view. The show features file tape of Alex speaking. "Look, there's Mom," Arlene says, "and Alex!" Griffin cocks his head.

"He's not been the same since Alex died," says Arlene sadly.

It's not as if Griffin lacks psittacine company. Another grey, Arthur (and like the young king, called Wart for short) joined the lab in 1999, mainly to work on computer-based tasks. Arthur doesn't say much, as he is developing other talents. Although he came to Irene at age one with a damaged right foot from an accident in chickhood, causing all four toes on that foot to face forward, Arthur is exceptionally good at manipulating equipment to solve problems. For instance, if presented with a long plastic chain at the end of which a nut dangles enticingly, Arthur has figured out how to pull the chain with his beak, anchor it with his foot, and pull again until the treat is within reach. (When Alex was faced with the same problem, he made no attempt to pull up the chain. Alex manipulated the world with language. His solution to the problem was to issue a command to his staff: "Pick up nut!")

Griffin doesn't have much to do with Arthur. In her memoir, Irene describes Arthur as "like the techno-freak teenager," absorbed in his toys. But Griffin didn't seem to be very good friends with Alex either. Irene had hoped that Alex would be a tutor to Griffin, but Alex refused to cooper-

Irene Pepperberg with Alex, Griffin, and Arthur

ate. In fact, Alex undermined Griffin's lessons when he could. When Griffin was being tested, Alex would shout out the answer before Griffin had a chance—or admonish "Say better!" or "Talk clearly!" Sometimes, like a naughty schoolboy, Alex would butt in to purposely give the wrong answer and confuse him.

"Griffin is the smart kid in the classroom who's shy and afraid he's not that smart. Alex was like the older brother who's always dominating you and whom you really resent . . . and now he's gone," Irene later tells me.

Only now, more than a year after Alex's death, has Griffin stopped looking toward Alex's empty cage, as if waiting to hear the older parrot's wisecracking answers. If he is grieving, is his grief tinged with relief? Regret? He can't tell us. Not even Alex had progressed to that level. "That's one of the worst things about losing Alex," Irene later tells me, over lunch, "not knowing what would have happened had we kept going with him. With Griffin, we're going back to kindergarten. We're learning our shapes. We're learning our colors." She sees a long road ahead of careful schooling and scientifically rigorous testing before he attains Alex's level of competency.

But Griffin may well one day far outdo his famous predecessor. Irene points out that for the first fifteen years of his life, Alex was an only parrot. "Griffin did not have that experience," she says, "nor has any other parrot in a laboratory, which might be why other labs have had less success." Alex, like an older brother, always outshone Griffin and often bullied him. "Griffin's only had a quarter of the training Alex did," Irene says. "Now Griffin's going to get that chance."

Though it can't be measured on a test, the day I leave the Brandeis lab I am certain of one thing: There's a lot more going on in Griffin's head than the color blue. Perhaps one day he will tell us about it.

The tales, in infinite variation, arose around the world, across ages, across cultures. Yet they all seem to make the same claim: birds, so many human traditions hold, taught people to talk, to sing, to dance.

The Fang-speaking people of Cameroon say African greys brought language to people as a gift from God. The Aruba tribesmen of Ghana also believe parrots taught people to speak. Hopi tradition tells us it was the mockingbird who brought the different languages to the different human tribes—and the Hopi still sing today what they tell us are the songs they learned from that original bird. The Quechua- and Shuar-speaking tribes of Brazil say that the macaws will remember the people's languages for them when the tribes themselves die out. And they may well be right. During his early-nineteenth-century explorations of South America, Alexander von Humboldt heard a parrot speaking a dead language. The parrot had been reared in a tribe that had been exterminated by foreign persecution and disease.

The ancient Roman poet-philosopher Lucretius felt that birds taught people not language, but music. And this they have certainly done. Beethoven's Sixth Symphony includes passages that imitate the cuckoo and nightingale and quail. French composer and ornithologist Olivier Messiaen considered birds the greatest musicians of all time. His seven books of *Catalogue d'oiseaux* are based on remembered and annotated birdsong. Mozart, too, may have learned from a bird. He had a pet starling whom he loved so much that when the bird died after three years in his home, Mozart wrote a poem in his honor and held a funeral for him attended by his closest friends. Starlings are great virtuosi, wonderful mimics as well as splendid singers, and incorporate so many different sounds into their songs that a flock of them is called a murmuration. Psychologists Meredith West and Andrew King, who study learning and birds, note that the beginning of the last movement in

Mozart's Piano Concerto in G Major is so like a starling's voice that the bird almost certainly whistled it; it is even possible, they say, that the bird actually *composed* it. Birds are still teaching people music today. American composer James Fassett has written a *Symphony of Birds*—consisting entirely of songs and calls of real birds.

The Kachin people of Burma hold that the birds taught people to dance—and even though the Kachin are now dispersed throughout Southeast Asia, they still reenact those ancient, bird-taught dances in a ritual called *manau* today.

These ancient stories speak a modern truth. Music and language may be among the crowning glories of human culture, but their twined evolutionary roots run far deeper than our short history. Of all the creatures to have arisen on earth, those whose capacities for speech and song seem to approach—or even exceed—our own are, surely, the birds.

Linguists have cataloged dozens of ways that human language, with its fabulously complex series of rules about things like conjugating tenses and use of reflexive verbs, differs from the communication systems of other animals. "Nobody's arguing that Alex learned human language," Irene stresses. She has been careful not to make that claim. "But those who study birds believe they have their own language," Irene says, "and it's very sophisticated."

Most birds, and particularly parrots, are difficult to study in the wild; they are shy, many of their interactions occur in trees where they are obscured by foliage, and they can fly out of sight in an instant. But it is known, for instance, that many species of parrots use special vocalizations reserved just for their mates; they are known to perform inter-pair duets that are quite different from the voices they use to communicate with other parrots in the flock. Thus, Irene writes in her book *The Alex Studies*, "the failure to document complex psittacine communication in the wild might be just that—a human failure, not a lack of capacity in parrots."

"We don't understand it yet because we don't have the tools," Irene tells me. "We've been studying birdsong for forty years, and we're just beginning to understand what's going on. People say, well, your birds know only a few elements of human language. But I say, how much do we understand of *their* system?"

Remarkably little. But scientists studying birdsong note its striking similarity to human language. Both are culturally transmitted communication systems. Both possess both syntax and structure—the order in which sounds appear dramatically affects how a song or a sentence will be understood. Both are learned in similar ways and affected by a similar gene.

Like human language, birdsong changes over time and space. Local song traditions change, as human language does. Largely for this reason, birdsong, like language, shows geographical variations or dialects. The song of the common yellowthroat, a rapid "witchity-witchity-witchity," gains more syllables in each "witchity" from north to south. Carolina wrens in Ohio sing faster than those in Florida. Bewick's wrens sing strikingly different songs in Arizona, California, and Colorado. Each valley of northern Scotland features a distinct chaffinch song. Black-capped chickadees who live on Martha's Vineyard have a distinct dialect found nowhere else. Elsewhere, these chickadees sing, "Hey, sweetie"—but on Martha's Vineyard they say it backward: "Sweetie, hey." Even though parrots are not songbirds, and technically their vocalizations are not known as songs, they show dialects, too: in some tropical areas, parrots of one species have such different dialects in different areas that people thought they were different species. The evolution of bird dialects is so like that in human language that Leiden University professors Carel ten Cate and A. Verhagen recently collaborated on a paper titled "Modeling Cultural Evolution: A Parallel Investigation of Changes in Bird Song and Human Language," based on how chaffinch song evolved along a chain of oceanic islands.

Birdsong and language also seem to be powerfully affected by nearly identical genes. The FOXP2 gene, which is shared by all animals as well as fungi, produces proteins that regulate other genes. But in both humans and songbirds, it is critical in the development of vocal communication. The only genetic basis of speech yet identified, the gene was first isolated in studies of a Pakistani family in which an inherited mutation of the gene caused afflicted members to be unable to learn to speak intelligibly, or to understand any but the simplest sentences. Later experiments showed that when the same gene was disabled in finches, the young birds suffered the same result: though they tried, they were unable to learn their species' song.

Young birds learn songs much like young humans learn language. As children first learn sounds, then words, then sentences, young birds first learn simple notes, then a song of "syllables" that make up "motifs" and then whole singing bouts, organized into correct patterns and timing. Ornithologists who study songbirds call the young birds' early babblings "subsong." In transforming subsong to real song, individuals not only imitate parents and neighbors but also improvise and invent: young birds transform themes, mix syllables, and incorporate new elements into their songs and may create individual signatures.

Children do the exact same thing. Sometimes they make up words. Sometimes they use real ones. Sometimes they mix them up together. No less a developmental authority than Piaget considered this practice integral to language development in humans. And though young children have ample opportunity to practice their early attempts at language with adults and other children, often they chatter to themselves, alone.

Practicing alone has several benefits. You don't have to worry about being embarrassed, misunderstood, or corrected when you make mistakes. You can play, uninterrupted, with a new sound until you get it right—or until you invent something you like better. You can explore new combinations and hear how they sound. And you don't have to worry if they're just nonsense.

Interestingly, this is precisely what Alex did when he was learning new words. A tape recorder hidden in the Pepperberg lab after the humans had left revealed that the parrot was playing with the new sounds he learned just like a child does.

Sometimes he would repeat his lesson verbatim. "How many?" Alex would ask himself. He could mimic the sound of the snaps Irene and the students played on the synthesizer. Then he would provide his own answer: "Foo-wah!"

Alex's private monologues increased soon after training began. As well as new words, his practice sessions also used phrases he had heard in speech unrelated to his formal training sessions: "You be good," "Gonna go eat lunch," "I'll be back tomorrow, bye." Sometimes he would intersperse these with the new words or sounds he was learning, perhaps as a sort of sound game. "Um" appeared in his monologues with increasing frequency as he was learning the word for "none": "Go on the gym um geepers," he said in one private practice session.

While learning the word "nail," Alex would make up words that rhymed with it: awail, banail, chial, mail, loobanail, blail, shail, lenail, jemail. Learning "green bean," he produced uninterrupted strings of sound including "keen green graeen," "bean green graeen bean," and "keen green bbbb." Alex was employing a universal human strategy: learning by embedding words in rhyme, meter, or melody. Most American children learn the alphabet by setting it to song; the ancients memorized epic history by setting the stories to poetry; athletes use music to coordinate their movements when they practice. (Oliver Sacks points out that our own language affirms the connections: rhyme and rhythm come from the same Greek root, carrying the meaning of measure, motion, and stream.) Perhaps this is why "green bean" was so easy to learn. Alex learned it within a week, while "nail" took nearly two weeks and "sack" never came. "Green bean" was as delicious to say as to eat. At times, he and Griffin would duet a two-word poem: "Green!" cried Alex. "Bean!" answered Griffin. "Green!" cried Alex again—and on it would go like a badminton match. The rhyme was so alluring that one human couple associated with the lab found themselves shrieking the parrots' duet at a waiter in a French restaurant.

No wonder both parrots and people love rhyme; rhyme is a sort of music, after all. In fact, many linguists and historians believe the first poems were not spoken, but sung. Perhaps the first human words were, too. Many believe that language itself arose from music. The kinship between the two is obvious: Plato observed that uplifting music resembles noble speech. Centuries later, Darwin proposed the idea that music was the immediate predecessor of language, arguing that the human voice was first employed to attract a mate. (That trick is still working; after all, by one calculation, about 40 percent of the lyrics of popular songs are still about sex and romance, and indeed, the sexual success of pop singers—and jazz musicians, and even nineteenth-century classical composers—seems to bear out the theory that a human's ability to make music, like a bird's ability to sing songs, can be powerfully seductive.) And, although the issue is hotly debated, a sizable number of experts think that our closest human relatives, the extinct Neanderthals, lacked complex language—but sang instead. In his book *The Singing Nean-*

derthals, University of Reading archaeologist Steven Mithen proposes these early people had a prelinguistic communication system more like music than speech. Only recently, says Mithen, have language and music, poetry, and dance separated into different ways of thinking.

Rhythm—the beating heart of Snowball's favorite dance songs, the rhymes in Alex's words—may be the mother of them all. Merlin Donald, the Canadian psychologist and cognitive neuroscientist, also believes humans existed without language for hundreds of thousands of years, communicating instead using the rhythms of gesture, posture, movement, and sound. Rhythm, he says, is not only a prerequisite of music and language, but the bedrock of human society, the foundation for everything from agricultural life to religious ritual behaviors.

Snowball might well appreciate that. Rhythm is certainly an important component of his world. To him, dancing is not merely a trick performed to please people. Just like Irene's audiotapes caught Alex talking to himself after she had left the lab, Irena's husband, Chuck, once walked in on Snowball while the five-disk CD player happened to be playing a German polka tune. The bird was dancing—all by himself.

As the late afternoon winter sun slants through the windows, Snowball is clattering his toys. The Schulzes, Gary, and I have devoured a vegetable platter and many cups of tea. The birds and we have enjoyed hours of our separate conversations. Now comes the moment I've been waiting for. I get to dance with Snowball.

Snowball steps onto Irena's hand for the short ride from his cage to the somewhat claw-battered grey upholstered swivel chair that is his favorite dancing platform. Recently she got a new desk. "But I realized I can't get a new chair," she tells me. "What if Snowball didn't like it?"

Snowball knows what's coming. His crest rises high, and his dark brown eyes shine with excitement. "Since it's your birthday," Irena offers me generously, "why don't you pick the first song?" I name one of my favorites: "The Lion Sleeps Tonight." Its tune is sung in a high-pitched falsetto (my own birds always seemed to enjoy the higher registers most); the song has a strong beat I think Snowball might enjoy; and it features a jungle theme to boot.

Chuck finds the song in an instant on the Internet: a recent YouTube video of the Tokens, the group whose cover of the originally Zulu song rose to number one on the Billboard Hot 100 in 1961.

Even before the jaunty chorus begins, the beat captures Snowball's imagination. Facing Irena and me, crest erect, he bobs his head enthusiastically and lifts his legs in time. *"In the jungle, the mighty jungle, the lion sleeps tonight . . ."* A moment after he starts, and without any prearranged plan, Irena and I spontaneously launch into the Pony. Anyone walking by and glancing through the windows to find two hopping, bopping, middle-aged ladies bouncing around in this suburban den this afternoon would surely conclude we look like loons—without even realizing we are rocking out with a parrot.

But it's irresistible. For me, this is pure joy: dancing, on my birthday, to my favorite song with my new friends—one of whom happens to be a cockatoo. My heart pounds, and it's not from exertion, but exhilaration. Researchers have found that heartbeats of finches quicken when they hear their own species' song; I imagine, perhaps, the finches are feeling a bit like I do now: Oh, *you* like that song, too! You must be one of *us*! Dance and song have powerfully served humans as a sort of social glue, binding groups together in synchrony, exhorting us to ecstasy. Why should we not share this bond with another species that, like us, finds that music stirs them to their very bones?

It is a sort of ecstasy I feel as I watch Snowball and find myself following his moves. Throughout the song, we watch each other closely. Snowball is devoted to Irena, and surely he enjoys dancing with her. But I am the new one, so he keeps his eyes on me. I don't think his movements are following mine, though. At first, at least, I have no doubt that it's me who's following *his* cues: I'm quite sure it's his idea to wave his left claw twice, then his right one once, while bobbing up and down and leaning slightly to one side. Because I am right-handed, and Snowball, like most parrots, is left-clawed, we make good partners. I feel he and I are caught in a sort of tractor beam of mutual attention, both dancing in perfect time, both with each other and with the music.

At the end of the first dance, the humans clap and call out, "Yay, Snowball!" Snowball bends his head and tosses his crest forward repeatedly as if taking a quick series of bows. "That was great. That was GREAT!" I cry to Irena and Chuck and Gary. "This is the best birthday dance ever!"

"Yes, that was really fun!" says Irena. "You were great together!" says Gary. "How about another?" asks Chuck. He next selects "Come Back to Me" by David Cook, another song with a strong beat, and one I do not know. I look to Snowball to lead, and I follow. There might be moments when I take over the beat or suggest, by a step or by my posture, a new move; I am not consciously keeping track, and I am not consciously trying to stay with his rhythm. Yet Snowball and I are in synch, remarkably so. How are we communicating this to each other? We seem to be keeping time with one another thanks to some mysterious sense akin to touch—yet we're not touching. It feels like the music is springing from inside us.

Finally, all three of us dance to no fewer than five different versions of "Everybody" that Ani Patel has created on the computer to test how Snowball would change his dance to altered speeds. I love dancing to the faster tempi. But with the slower ones, I have the same problem as Snowball: it's very hard to stay with the beat, and it takes a few missteps before we both get in synch.

The only reason we stop at all is that I don't want to tire Snowball out and risk prompting a tantrum or seizure. I feel I could dance with him into the night.

I'm not a great dancer, but I have danced a lot—from college dance marathons to Amazon village fiestas to thrice-weekly step aerobics classes for the past twenty years—and only once before do I remember synchronizing with a partner so well. That was with my father, and we were playing tennis, not dancing. I was a teenager. One day when we were playing, we found that I absolutely knew where he was going to hit the ball next, and he knew exactly where I would return it. We found this was more fun than trying to outwit each other. My father was my hero; I had watched and emulated him every day of my life. I shared half of his genes. But with Snowball, I had no such history. Snowball and I had just met, and besides, he was a cockatoo—as distantly related to me as a dinosaur.

Because I like Snowball so much, I realize it is possible I only *thought* I synched so well with him on all but the slowest tempi. But no; I did not imagine this. As it turns out, Chuck recorded a short video of us dancing on his cellular phone. He later e-mailed it to me. Interestingly, the song he taped was the second one, the one I had never heard before, the one I would have normally danced to most clumsily. And yet in the video, I am not clumsy. I am taking my cues from a virtuoso, from someone whose lineage, the ancients say, taught ours. Everyone who views the video, as well as Gary, the most objective scientist I know, who saw it actually happening, agrees: although we are separated by 325 million years of evolution, Snowball and I move together, as if in a mirror.

Crows

Birds Are Everywhere

A nightly spectacle begins in Auburn, New York, every fall and winter day, as dusk approaches. First, lines of crows start flying in from three or four different directions, cawing, as the big black birds begin to mass for their nightly roost. They start to coat the naked trees like leaves—big, dark, moving, noisy leaves. Separate lines of crows converge until they become a storm of crows, an explosion of crows, and finally a great flowing river of crows. The air is alive with hurrying black wings swirling down from the sky.

"Oh!" I gasp, and my hand flies to my heart. Though I am standing at a city dump watching one of the commonest species of birds on earth, I could not be more thrilled were I surrounded by a host of angels.

I am witnessing a natural phenomenon that I had always hoped to see: the massing of crows into gigantic, communal winter roosts. Though sometimes groups of crows start to gather into night roosts as early as August, only as winter begins in earnest do they mass by the thousands, and sometimes by the hundreds of thousands. A roost near Washington, D.C., was estimated to contain 200,000 birds. One in Rockford, Illinois, hosted 328,000. A roost along the Delaware River in Pennsylvania attracted 500,000. One at Ft. Cobb, Oklahoma, had more than 2 million. The crow roost where I stand transfixed is much smaller, but still impressive. Here in Auburn, a small city in upstate New York, the crows outnumber the human population by nearly two to one.

Crow roosts like this have intrigued naturalists for centuries. Recounting his vivid observations of 12,000 crows returning to a winter roost near Ipswich Beach, Massachusetts, in 1918, Charles W. Townsend wrote he "wished for eyes all about the head, sharpened wits, and a trained assistant to take down notes." I know how he felt: a gathering this large of living creatures seems too big for our limited senses to take in, too grand for a mere human to process. Other birds, too, form large, impressive roosts, especially red-winged blackbirds and starlings: on fall and winter afternoons and evenings, their fluid, fast-moving flocks swoop through the skies in cloud-like formations so spectacular that birders post their locations on websites like trails.com in the United States and wildaboutbritain.co.uk in the United Kingdom. Their splendid profusion echoes the wondrous swarms of migratory monarch butterflies that cover whole trees in the forests of Michoacán, Mexico; the olive ridley sea turtles who come ashore by the millions on one night each year to lay eggs on the shores of Orissa, India, in their annual arribada. Gathered together, so many individuals seem to form, for a moment, a sort of "superorganism," as E. O. Wilson calls a colony of social insects; all seem to be expressions of a single idea, the will of one Supreme Being, the waking dream of one great, collective Mind. A videographer who posted a YouTube video of flocking starlings in Northampton, England, must have felt the same way, setting the flock's shifting shapes to the haunting music of Norman Greenbaum's rock classic "Spirit in the Sky."

The crows may not be of one mind, but certainly they have a plan. They aren't staying overnight in these trees. The dump is at the outskirts of the city. The crows—many, if not most of them—are using the dump as a staging area in which to meet in order to fly on to their next destination. For more than one hundred years, Auburn's crows roosted in the woods outlying its farms and suburbs. But these days, things are changing. Now the crows are headed downtown, into the heart of Auburn—a city of fine old churches, two-storey wooden houses, pretty little parks, 28,000 people, and from October to March, a city of 50,000 crows.

My friend, the poet Howard Nelson, wrote a poem about it, titled "Kingdom of Crows":

> *Auburn has been blessed, or cursed, with not just a flock of crows, or a*
> > *bevy,*
> *but with a hoard, a plethora, a multitude of crows,*
> *a throng, a legion, a host of crows,*
> *a swarm, a mob, a rabble of crows,*
> *a very big bunch of crows,*
> *a murder of crows (should we murder the crows?),*
> *a batch, a swad, a gang, a crush of crows,*
> *oh, definitely a large crowd of crows . . .*

I learned about Auburn's crow roost from Howard, who teaches at nearby Cayuga Community College. He's written several poems about the crow roost; he's fascinated with the gathering. In our letters back and forth we have mused, as generations of naturalists have before us, about why the birds mass like this. No one really knows. Do they congregate for warmth? For protection against predators? Are they drawn by nearby food? Do they gather to exchange information? Of one thing we're certain: whatever their reasons, they must be excellent—for crows are uncannily intelligent birds. The authors of *Aesop's Fables* knew this well: in one of the stories, a thirsty crow finds a pitcher with water in the bottom, too far for the bird to reach; he gathers pebbles and drops them in the pitcher until the level of water mounts high enough for him to drink. It's the sort of thing a crow really might do. Crows are experts at figuring out problems. As Henry Ward Beecher, the nineteenth-century American preacher and lecturer (and brother of Harriet Beecher Stowe) said, "Even if human beings wore wings and feathers, very few would be intelligent enough to be crows."

Over the past several years, Howard has invited me again and again to come see the roost. And then one day he called to say that the crows were going to be driven out of the city.

I got in the car and headed west to Auburn.

* * *

I have always loved crows. I am grateful to them. They have probably saved the lives of many of my hens. Bravely risking (and sometimes losing) their lives, groups of crows mob day-flying predators like hawks, chasing away and cawing loudly; my hens overhear and take cover. Mobbing crows have led me to day-roosting owls (seeing an owl, it is said, is a portent of life change: your life has already changed, I say, and for the better, for you have seen an owl!). Some of the friends I love best have raised and released orphaned crows. My friend and photographer Dianne Taylor-Snow raised one, named Crow Baby, who cuddled and slept with her and her husband. He loved to raid her jewelry box for shiny objects. Though she released him to the wild decades ago, when she pulls out a book from her home library, often she finds a long-lost ring or necklace or a piece of tin foil stuffed between its pages. She never fails to smile.

One of our neighbors feeds our local crows, once offering them an entire freezer-burned turkey. (Our dog Sally discovered this bonanza one day, to her delight.) One friend had crows as babysitters. Ben Kilham, today a wildlife rehabilitator who specializes in raising and releasing orphaned bear cubs, remembers a childhood spent with a leopard roaming the living room and a crocodile in the shower. His parents, both doctors, adored animals and rescued those in need; accomplished bird-watchers, they adopted two orphaned baby crows who grew up as part of the family. The crows would play with Ben in his playpen, perching on the rail. Ben used to play with a silver bell hanging from a string. When the infant let the toy go, one of the crows would seize the bell and the baby would try to get it back. The crows never so much as pecked the boy, as his dad, my friend Lawrence, wrote in his book *The American Crow and the Common Raven*—not even when Ben would grab at their wings or tails. After all, they were taking care of baby Ben just as they would have done for nestlings in their own family.

Like us, crows are long-lived, with lasting family ties and long memories; a lucky crow can live to twenty, a raven, the crow's closest kin, to forty. Like us, they live in close-knit, stable family groups, which is why you almost never see a crow alone. If you do, look further: you are probably seeing a sentinel, a member of the family whose job it is to keep watch for danger. Look carefully and you'll see the other crows: in the warmer seasons, usually a group of three to five—mom, dad, and at least one bird from a previous year staying around to help his parents raise the current brood of youngsters. In the spring, you'll usually see three or more crows building the same nest. A young crow might spend up to six years with his parents before leaving to breed on his own. In her studies of crows in Stillwater, Oklahoma, Carolee Caffrey, then of Oklahoma State University, found that adult crows sometimes visit their siblings living elsewhere, as we do; some even move in with their siblings' families; some set up their adult breeding territories adjacent to their parents' when they grow up. "Extended families of at least three generations would sometimes spend time together," she found.

As with people and wolves, "one of the most difficult of all things for a crow to endure," wrote Larry Kilham, "is to feel alone and separated from one's own kind." Bird rehabilitators who care for crows never release a crow alone; a lone crow is a dead crow. They need one another, as we do—not only for company, but for survival. Crow families cooperate to hunt and gather, defend their territories, and care for the young. While studying crows in New Hampshire and Florida, Larry Kilham saw groups of crows working together to raid food from otters, vultures, and waterbirds. One technique is for the crow to pull the feasting animal's tail, distracting it long enough for a partner to make off with the booty.

"Crows," Larry wrote, "seem, as though by convergent evolution, to have something in their psyches corresponding to something in our own." It seems to me in fact that they reflect the best of the human psyche. Crows love to play: in the winter, I have seen them slide down snowbanks like otters; they also love to drop and catch found objects like toys and play games of tug-of-war and catch-me-if-you-can with one another as well as other species. Crows have a sense of humor: Larry watched crows tweak the ears and tails of other animals for pure mischief, land on sleeping cows and sheep to startle them awake, and fly off with unattended car keys as well as clothing drying on clotheslines just to watch people run after them.

Crows are loyal. They'll feed sick family members and will guard them until they get better; injured crows call to flock-mates, who rush to help and fend off predators. And crows are smart: McGill University animal behaviorist Louis Lefebvre, inventor of the world's only comprehensive avian IQ index, ranks the crow family, the corvids (including about 120 species of crows, ravens, rooks, choughs, nutcrackers, jays, magpies, tree pies, and the

African piapiacs), at the top of the list. Konrad Lorenz considered the crow's close relative, the raven, as having "the highest mental development" of any bird; even Irene Pepperberg admits crows and ravens might even be as smart as her parrot, Alex.

So why would anyone want to get rid of them?

* * *

A crow
has settled on a bare branch—
autumn evening.

—Basho

50,000 crows
have settled on bare branches—
big mess in Auburn.

—Howard Nelson

This is the crux of the problem. Auburn mayor Tim Lattimore tells me that at the bank, the tellers leave work each night beneath umbrellas and dash to their cars. At the YMCA, beneath some of the crows' favorite roosting trees, patrons exiting at dusk are often hit several times. At the historic Seward House, the gutters are so plugged with crow excrement that the new $400,000 roof has started leaking on the antiques inside. When people walk into city hall, they track in crow droppings from their shoes onto the carpet. Folks won't feed the city's parking meters because they're too splattered to touch. "I've been trying to attract Fortune 500 companies here," the mayor says with exasperation, "and I can't have Fortune 500 companies visiting us when they have to walk through fecal matter to get to the building. It doesn't give the right presentation we want in the city of Auburn." When he took office, downtown Auburn's crow roost was the largest in New York State— hardly the Chamber of Commerce boast he had hoped for.

The crows are the talk of the town. Some folks, like Howard, welcome the drama of their dusky arrival. Others, intimidated by so many noisy, flying birds, consider them a nightmare right out of an Alfred Hitchcock film. But everyone agrees on one thing: the droppings beneath a roost of fifty thousand crows can be quite intimidating. The only business that considers this a bonanza is the local carwash.

Crow roost at Troy, New York

Auburn is not alone in this problem. Nearby Troy, New York, has a crow roost of twenty thousand, directly above, among other things, a federally subsidized apartment building and Gurley Precision Instruments—which just finished a major renovation to its historic building and installed a new blacktop for the two hundred employees who park there. The city of Utica has a roost of ten thousand that town fathers would like to see gone. The state capital, Albany, is also plagued by crow droppings from its large roost. Of course, people produce far more waste than the crows, but it doesn't rain down from the sky. (But even worse, we end up drinking it: antibiotic and hormone drugs in human urine get flushed into inadequate wastewater treatment plants and then find their way back to the human water supply.)

New York State reflects a nationwide and possibly a worldwide trend. University of California biologist W. Paul Gorenzel and four colleagues conducted surveys of U.S. Fish and Wildlife Service state directors in thirty-nine states, and one of officials in 473 towns and twenty-nine counties in California. In the national survey, twenty-seven of the thirty-nine directors identified eighty-six urban crow roosts; in California, fifty-seven cities reported crow problems, twenty-four of them because of winter roosts in downtowns. "Results suggest crow populations are increasing, a shift to urban roosts has occurred and is still in progress," the biologists conclude, "and problems with urban crow roosts are likely to increase."

From Vancouver, Canada, to the coastal cities of East Africa (where Asian house crows were introduced in the nineteenth century), residents complain of crow population explosions. In compulsively clean Singapore, the National Environment Agency now considers city crows "a public health nuisance." In Japanese cities, where government officials say the population of their native jungle crow has grown enormously since the 1990s, the birds cause power outages; the crafty birds incorporate the city's fiberoptic cables into their sturdy stick nests. Over the short span of just two years, crows caused fourteen hundred blackouts in Tokyo—one of which shut down a segment of the city's famous bullet train service.

Why are crows moving into cities? "It's some sort of cultural revolution in Crowdom," Cornell University Laboratory of Ornithology crow expert Kevin McGowan tells me. "Exactly why it's happening, we don't know. But we definitely are experiencing it. The crows are changing their behavior within our lifetime, and whatever the reason, it's spreading all over." Part of the reason may be that there are simply more crows. Despite an epidemic of West Nile virus not long ago, data from the North American Breeding Bird Survey and Audubon Christmas Bird Count show American crow populations have been growing at a rate of about 1 percent a year since 1966. Crows fall into a category of wildlife that biologists call "invigorated species." While all too many populations of wild creatures are shrinking due to human activity, a few benefit. Crows, like raccoons, Canada geese, coyotes, and white-tailed deer, have not only adapted to but have taken advantage of some of the changes humans have made to the landscape—and one of the major changes we have wrought has been to create cities, where, only in the past ten years, for the first time in human history, the majority of humans on the planet have come to live.

Crows may flock to the cities for many of the same reasons we do. Cities offer many amenities. For instance, the fine dining: Auburn's sanitation workers swear to me that the crows know the garbage routes better than they. Human refuse features a smorgasbord of pleasure for a winged scavenger with an eclectic palate. The growing abundance of urban garbage, since Japan adopted the wasteful Western lifestyle, is credited for Japan's crow population explosion. (Alas for the crows, though, this diet of literal junk food isn't any better for them than it is for us; bird rehabbers who treat urban crows tell me they see many signs of poor nutrition, including badly developed feathers.) Cities tend to be warmer than suburban or rural areas, because cement and asphalt retain heat. Crows may also appreciate the bright lights of the big city: it's easier to spy a night-flying predator like their worst enemy, the great horned owl. And as for the presence of all those city people, crows don't mind us at all. Though crows were once shy of humans, when every day was open season on crows, the birds have taken notice of laws that restrict bird hunting to certain time periods. And they have also noticed that hunters tend not to discharge their weapons downtown.

"The living here is pretty good," Richard Chipman tells me, "and the crows know it." His job is to convince them otherwise. Genial and soft-spoken, Rich is here in Auburn leading a team of federal scientists and city employees who are working to drive the crows out of the city—without hurting them. He has invited Howard and me, as well as a radio producer with the show I sometimes work with, to come with him in one of the four U.S. Department of Agriculture trucks on what feels like a cross between a police patrol and an urban safari. We're searching out the crows' roosts, and then, like cops on the beat facing loiterers, urging them to move along.

We're an eclectic group: Rich, the crow specialist, blond and blue eyed with a blue baseball cap and orange USDA jacket. Howard, the poet, slender, quiet, and graceful, with thick grey hair, a greying goatee, and soulful eyes like a mountain gorilla. Our driver is Dave "Spider" Ganey, a strong, tough guy with a dark moustache who has worked as a heavy-equipment operator for the city for thirty-five years, mainly cutting down trees for a living. My producer is a chatty, urban, middle-aged *Sex and the City* type who looks like she'd be mortified to get crow shit on her shoes. And me. I don't mention to her that, while raising baby chicks in my office, I not infrequently have bird shit in my hair.

Rich stays in touch with the other three teams by cell phone. The ringer sounds. "Hey, Adam, did you find any?" Rich says. He mouths the answer to us: *Yeah—big-time!* "Here, let me give you to Spider." He does. "Van Anden

Street?" says Spider. "Give us two minutes. We'll be there, champ." We rush to join the other team at the scene: a residential neighborhood with tall pines and oaks, not far from a ball field. As soon as we get out of the truck, we hear the crows cawing as eight or nine hundred birds jostle for position in the trees, trying to settle in for the night. But Rich is here to make sure they can't.

What Rich is directing is known as a hazing program. Basically, his team is being paid to irritate the crows with pranks that sound worthy of a fraternity initiation. Spider is the first to deploy the device of choice in this neighborhood: the thousand-dollar Avian Dissuader laser. He points a beam of bright red light into the trees, wiggling it around. It doesn't have any effect on deer or other mammals, but birds just can't stand it. The wavelength is carefully set not to damage birds' sensitive eyes, but it seems supremely irritating. Who could think of sleeping under such conditions? Crows explode from the branches, cawing. They're warning crows everywhere: "This is a bad place! Let's go away!"

In fact, on the opposite side of the street, the other team is saying this, too. One of the men holds a megaphone aloft. It is broadcasting an amplified recorded crow alarm call. Crows have many different known calls with various meanings, some of which are understood by humans, and some of which are not. A slow barking caw—when a predator is spotted at a safe distance—calls other crows to fly in and investigate. Loud, angry caws bring birds in fast to mob a hawk or owl at close range. Hunters use recordings of a dying crow's heartbreaking calls for help to attract other crows, who come to the rescue of a fallen comrade only to be killed themselves. Different researchers classify different calls differently, and no one claims to know what they all mean: some are called "woo-ah" notes, "wow-wow" notes, organ notes, coo notes, rattles, clicks, and chirps. Also, crows and their kin are gifted mimics, who can speak if they choose, like parrots (even though the crow and parrot families are not closely related). Some pet crows have learned more than one hundred words and up to fifty complete sentences. British ornithologist Sylvia Bruce Wilmore tells of a Staten Island crow whose owner trained him to pick pockets; if the bird found no money, he would scream "Go to hell!" and fly away. Konrad Lorenz's raven, Roah, could speak his own name and used it as a call note for Lorenz, uttered in Lorenz's voice. Crows, as well as their relatives, ravens, rooks, and pinyon jays, develop individual, specific signatures to their calls, like a dolphin's signature whistle or a person's name. As dolphins do, when a crow calls a missing friend or mate, he will use this specific sound to call him by name. If the other crow is free to do so, she flies immediately home.

But this broadcast, recorded call is a rapid, high-pitched, staccato message. It clearly signals retreat. It has the effect of someone yelling "FIRE!" The crows depart in a panic, urging their fellows to follow.

I confess to a twinge of species guilt while watching this, as I do when a running tot scatters a flock of peacefully feeding pigeons. Why can't we just leave other species in peace? But Rich is delighted, and for all the right reasons. "They're a very intelligent species," he tells me. "It doesn't take them long to learn. They're going to move on. This is great. This is what we want."

The news crows convey to their fellows can profoundly affect the whole crow community's behavior. This was powerfully documented in studies conducted by Dr. John Marzluff, professor of wildlife science at University of Washington in Seattle. He and his students capture, tag, and measure wild crows for their studies, a process that involves shooting a net over a crow and briefly pinning it to the ground. The crow's ordeal is over in less than ten minutes, but apparently it's a bitter and memorable experience. Marzluff soon noticed that crows around campus grew wary of him. He wanted to test whether they recognized his face. In his experiment, whenever he or his students trapped crows, they wore a caveman mask; at other times, when they were observing the birds but not attempting to catch them, they donned a mask of the unpopular U.S. vice president, Dick Cheney. From the moment the first crow was captured, campus crows consistently mobbed the person wearing the caveman mask, screaming and chasing him and trying to drive away what they clearly saw as a dangerous intruder. The misdeeds of the former vice president notwithstanding, the crows left the person wearing the Dick Cheney mask alone. But the most surprising aspect of the experiment was this: each time the caveman mask appeared, the crowds of mobbing, screaming birds grew larger. "Now on campus, almost every bird you see will scold us when we wear that mask," Marzluff told an interviewer shortly after publication of his book *In the Company of Crows and Ravens*. "We only caught seven individuals with the caveman mask. And that was two and a half years ago. That's a long time." Next he plans to test to see whether the tradition has spread. Will young crows raised on campus but now living elsewhere still mob a person wearing that mask? He'll be taking the caveman mask up and down the Puget Sound area to see.

But a caveman on campus is nothing compared with the shock Rich is planning next for the crows of Auburn. As we pull the USDA truck over near a small urban park, named after the former slave and abolitionist Harriet Tubman, I hear Spider inquire, "Do you have your artillery rounds, Rich?"

Artillery?! To my horror, Rich is taking out a gun. But it's a cap gun, from which he is about to launch screaming fireworks into the night skies above the roosting crows. "People aren't happy with exploding fireworks in their neighborhood during the dinner hours," Rich notes. "And crows don't like them, either, in their bedroom at night."

The fireworks zigzag into the night sky, shedding a stream of orange sparks and shrieking like a hundred-decibel siren. This particular model is called the Screamer Siren. Another kind is called the Bird Banger, which issues a loud, time-delayed charge as it explodes one hundred yards downrange. The birds respond with their own explosion. As thousands of crows flee the trees, their wingbeats sound like a rainstorm. Their deep black forms against the grey city sky look like night itself taking flight. Their caws of dismay spread the news all over town: there goes the neighborhood.

* * *

> *. . . a brigade, a battalion, a detachment, a fleet of crows,*
> *a conclave, a congress, a convocation, a convention,*
> *a vast cawing caucus of crows,*
> *a symposium, a get-together, a mass-meeting of crows,*
> *a synod, a plenum, a quorum,*
> *a conglomeration, an accumulation, a medley of crows,*
> *a fardel of crows, a truss of crows, a seroon of crows . . .*

Tom Lennox is no poet, but if he were, he might add this line to Howard's poem:

> *. . . too damn many crows.*

"They're like rats with feathers!" the tall fifty-three-year-old in the cow-boy hat tells me over at the J & B Bar and Grill. Crows, to Tom's mind, are good for only one thing: target practice. He and his hunting buddies have made this bar the official headquarters for the annual Crow Shoot, sched-uled for the second weekend in February. "I'm taking a negative and turning it into a positive," he insists. After all, shooting crows is perfectly legal in winter in rural New York, and Tom and his buddies have been doing it for years. But these days, they've turned it into a public event, kind of like a win-ter festival. The weekend kicks off with the official Crowmobile: a custom-ized 1986 Pontiac equipped with flapping wings that proceeds down the main street. All the while the vehicle excretes lifelike giant droppings from the back (the secret ingredients: Cremora and windshield wiper fluid). Dur-ing the weekend, dozens of four-member teams, each paying a registration fee for the privilege of joining the hunt, work together to shoot and collect as many slain crows as they can. The team that kills the most wins the grand prize of four hundred dollars. Last year, Tom brags, the hunters bagged more than one thousand crows over the course of the weekend.

Tom styles himself a bit of a redneck for the purposes of our inter-view. "Super Bowl's over with . . . NASCAR doesn't start for another week . . . what are people like me supposed to do that weekend in February?" he asks, and laughs raucously. Later we meet for breakfast. He's not a mean guy. He's funny and honest; he's a smart businessman who runs a successful tree-removal service for a living. And to my surprise, though his views of crows shock and upset me, I find I quite like him. Tom has a tender heart. From the registration fee charged each hunter at last year's shoot and sales of Crow Shoot T-shirts and memorabilia, he and his buddies were able to raise a thousand dollars, which he donated to a local citizen fighting a rare cancer. And his kindness is not reserved for only humans. Tom and his wife also rescued and adopted an abused dog who had been tied up and neglected in her previous owners' backyard. Now the sweet, grateful dog travels with Tom almost everywhere.

But crows, to him, are another matter. "They're like, taking a septic tank and pumping it into the street!" he says with disgust. "Imagine you had thousands of guys urinating into the city drinking water! Each crow prob-ably does a dollar's worth of damage a day," he insists. "They ravage songbird nests. They get into the duck eggs. They tear off the farmers' silage wraps. They're overwhelming everything!"

Tom's not alone in his feeling. Some people just hate crows—especially when there are a lot of them. Through the centuries, many people have found members of the crow family "as unappealing as cockroaches and as undeserving of sympathy," writes Candace Savage in her book on crow intelligence, *Bird Brains*. While most people can summon admiration for some of the more colorful members of the family Corvidae—like the familiar blue jay and the stunningly iridescent, blue-headed green jay of Mexico and South America—the majority of the family is black, and these species elicit the same suspicions as do black cats, black sheep, and black hats. The prejudice is reflected in our language: after all, a large group of crows is called a murder; a flock of ravens, an unkindness. The same sentiment is reflected in art: American realist landscape painter Winslow Homer's iconic 1893 *Fox Hunt* depicts the popular nineteenth-century notion of crows as symbols of doom. In the painting, two low-flying crows harass a red fox as he makes his way over a snowy landscape, while in the background more crows lurk ominously. In life, corvids and canids often team up as hunting partners—ravens lead wolves to prey, which the wolves can open with their teeth so the ravens can partake. But in Homer's painting, the crows are chasing and frightening the fox, and the viewer wants to shoo the birds away. And despite the fact that the most famous quote of his writing career is attributed to a raven, even Edgar Allan Poe considered the whole crow family "grim, ungainly, ghastly, gaunt, and ominous." As we humans do, crows and ravens eat carcasses, though they don't get theirs at the supermarket; farmers are incensed when the birds feed on dead and dying livestock (animals who, if healthy, the farmers would kill and eat themselves). Around the world, crows are accused of harassing livestock, raiding crops, and spreading garbage (who put the garbage there in the first place?). In 1989, the British House of Lords rose in outrage against a proposal that corvids should receive some sort of protection, like other birds. One lawmaker cried out in reply, "Capital punishment for the thieving and murderous magpie!" (But what if all the ravens—fellow corvids—left the Tower of London? Legend warns this would spell the fall of the Kingdom, and to prevent such a catastrophe, the nation employs a royal raven keeper.)

Even bird lovers hate crows. My friend, the big-hearted hummingbird rehabilitator Brenda Sherburn, shoos them away from her feeders. "Go away! Get out of here!" she yells at them. She wishes the rehab center she works with wouldn't treat their injuries or rescue their orphans. My falconry instructor Nancy Cowan doesn't like them either. Crows will go after her hawks and can kill them. "There goes trouble," Nancy says whenever she sees one. "They're a menace." When she was giving a falconry demonstration at a local school with her bird Indy, the instant she saw three crows fly overhead, she knew she'd soon have to cut the program short. "Those crows saw my hawk," she told the principal, "and they're going to tell their buddies to drive my hawk out of the area." Within five minutes, she said, "Every tree in the schoolyard was covered in crows—and Indy knew they were there for him." Six hundred cawing crows drowned out Nancy's voice as she tried to speak to the schoolchildren, and the crows had the big hawk so intimidated he would fly no higher than ten feet. She packed up and left.

But as one of the world's top crow researchers, Cornell's Kevin McGowan, points out, "These birds aren't a gang of nasty villains. These birds are just birds. American crows are among the most family-oriented birds in the world!" But they do suffer from a PR problem, exacerbated by the fact that they feed on the corpses of farm animals and, especially on the battlefield, people. Because of this, crows and their relatives have been associated in much of European mythology with cruelty, death, and disease.

"Even in these large flocks, crows don't represent a health hazard at all," McGowan stresses. They do not spread disease to people—not West Nile virus (it's transmitted by mosquitoes, not birds, and afflicts far more robins than crows) and not Asian bird flu (which is spread from birds to people by domestic fowl only, and usually only after very intimate contact with them— one documented case occurred after a man used his mouth to suck the mucus from the nostrils of his sick fighting cock, who was losing a high-stakes battle in the ring). Crows' impact on other songbirds' populations is small— minuscule, in fact, compared to that of house cats (and cats' depredations are themselves small compared with human overpopulation and its attendant deforestation, pollution, and climate change). Crows do more good than harm to human food crops. In her book, Savage cites a study from New York State that found only 1 percent of crows' summer diet was field corn.

Another study found that a single family of crows will devour forty thousand grubs, caterpillars, and other crop pests during their nesting period.

So why then do so many people, even smart, kind people like Tom, persist in their hatred of crows? Possibly it's just that the birds are big, noisy, and numerous. But I think it might be more than that. Years ago, I spoke with an environmental studies student at San José State University who was working on a paper for a class called Sociology of the Environment. "I think it has a lot to do with our feelings about cities in general," this thoughtful young woman told me. "When animals can manage to live with us, we hate them for that. We consider them less than animals. But what does that say about us? That any animal that chooses to live with us can't be any good?"

Perhaps, for some people, crows are too close to us for comfort. As crows move into cities, not only are they getting closer to us in physical proximity; eerily, these winged aliens, with horny beaks instead of lips, scaly reptile feet, superhero vision, and the ability to fly, are coming, in other ways, to resemble us more and more.

For instance, crows use tools. Until Jane Goodall discovered chimps using twigs to probe for tasty termites, tool use had been considered, like language, an uncrossable boundary separating humankind from the rest of animate creation. But in the wild, in the laboratory, and in the city, crows and their relatives turn out to be expert tool users. Wild New Caledonian crows of the South Pacific not only use tools, and not only make tools, but will use two different tools in succession if they deem it necessary to accomplish their goal. The birds were caught on tiny, bird-borne video cameras using their bills to whittle twigs into hooks and tearing leaves into barbed probes. Sometimes they used one after another to fish a particular bug from a crevice.

In the city, crows go even further: they manage to use human tools to their ends. They even get us to do most of the work for them. John Marzluff reports that in Japan, crows have figured out a way to exploit cars and traffic lights to take advantage of a new food source they never ate before.

Walnuts are a crop new to Japan, but lately groves seem to be springing up everywhere. Crows find walnuts tasty and nutritious, but the shells are hard to open. The solution: crows pluck the nuts from the trees, then fly to perch on the traffic signal at the nearest traffic intersection. When the light is red, they fly down and place the nuts in front of waiting cars. When the light turns green, the cars run them over, cracking the hard shells. When the light turns red again and the cars stop, the crows fly down to safely eat the nutmeats. "Some people in Tokyo are very careful to make sure they run over the nut for them," Marzluff adds.

American humorist Ian Frazier has observed that if crows were part of the human, corporate world, their success story would be widely admired; others would strive to emulate them. He makes light of this in a hilarious essay, in which he plays a marketing director who has just met with his new clients, who happen to be the crows. He has just come up with a new slogan for them: "Crows: We Want to Be Your Only Bird." They probably do.

* * *

> *. . . a drove, a school, a snowdrift of crows,*
> *an omnium gatherum of crows,*
> *a galaxy of crows, a séance of crows,*
> *a rendez-vous*
> *of crows.*

On the other side of Auburn from the J & B Bar and Grill, earlier this fall, the crowds of crows drew quite a different crowd of humans. Howard read this poem as part of a series of lectures and readings dedicated to crows at the Schweinfurth Memorial Art Center. There was also an exhibit of paintings and sculptures of crows. It was the sort of thing Rita Sarnicola thinks could really put Auburn on the map—if only the crows were properly appreciated.

She's one of the founders of CROW: Citizens Respectful of Wildlife. She's appalled by the Crow Shoot. She's saddened by the hazing efforts. She's delighted that the crows roost in Auburn's downtown and wishes they could stay. "The potential is just enormous," she tells me. "The crows would draw tourists. They could be the focus of science studies. So many people love nature watching. I wish the mayor and the city would understand the potential! It would bring in tourists and stimulate a whole lot of business." She imagines expanding the crows-in-the-arts theme into a festival for both adults and children, with art and poetry contests for kids and opportunities for humane education.

Josh Klein, while a graduate student at New York University, had an even rosier vision. He considers crows "synanthropes," sophisticated urbanites who have so seamlessly adapted to human cityscapes that they own it just as much as we do. As a master's thesis project, Klein built vending machines for crows. He offered a flock of crows in Binghamton, New York, coins and peanuts from a dish attached to his machine. Then he took the peanuts away. When the birds searched the dish for missing nuts, they ended up pushing coins into a slot, causing more peanuts to appear. The birds soon figured out the system. Soon the whole flock was out searching the city for loose coins. Klein's project provides a glimpse at what he hopes might be possible if humans collaborate with wild creatures in the city. To this end, Klein created the Synanthropy Foundation and is working with students at Cornell and Binghamton universities to see if such techniques could reward crows, pigeons, or rats for performing chores for us like sorting our garbage.

Meet the neighbors: crows are smart, savvy, and increasingly urban.

Perhaps, one day, crows will be integrated in this way as part of human society. Or—the same possibility, but seen from a crow's-eye view—we into theirs. For the moment, though, the crows of downtown Auburn are flying away. By the time I drive back to New Hampshire, Rich and his USDA team have completed five days of hazing. The program looks to be a success. No longer do they roost by the park where we shot off the screaming fireworks. No longer do they drip their excrement on the historic Seward House, or the bank tellers, or the patrons of the YMCA. Where have they all gone? We don't know. When we stop to scan the big trees near the state prison, Spider tells us that last week we might have spotted nearly a thousand crows roosting here. Now we see only one, a sentinel, silhouetted on an outermost branch against the grey city sky. "Caw! Caw!" he cries in high-pitched, staccato alarm, and then he and his family fly out of sight.

It's been three years since I visited the crows of Auburn. It is Crow City no more. The USDA program worked brilliantly. Howard tells me there are still plenty of crows around; they're just not roosting downtown. Quite a few now roost in the trees near his rural house in neighboring Moravia.

Some folks might say the crows—smart birds!—learned their lesson well. And that is true. But the crows taught something to the people, too, as Spider told me after our first night on crow patrol together. "Like a lot of people, I think, I never gave crows much of a thought before this project," Spider admitted. "I guess, like a lot of people, I probably thought they were just flying rats."

But then, the first day Rich took him out, they found the gigantic staging area down by the dump—the same place I first saw them. And though Spider is no bird-watcher, his reaction was the same as mine. "We got them moving and it was like, holy mackerel!" Spider told me. "I thought, this is incredible! All of a sudden, like twenty thousand or more going up at a time—a gigantic black explosion! The noise, the beauty of the thing . . . If that was on TV, I'd watch that. I'd love that."

Spending time with the crows changed everything, he told me. "I find them incredibly intelligent, versatile. I'm fascinated by them. This has been a great learning experience." Spider, who's spent thirty-five years cutting down trees and moving heavy equipment, was never one to worry about the comfort of birds. But he told me how very glad he was that not one crow was hurt in the hazing operation. "I appreciate them a lot more than I ever did before," he said.

A number of blessings, it seems, came out of the crows' brief stay in Auburn's downtown. And here's one more: another poem Howard wrote, one of my favorites. It's titled, simply, "The Crows." This is its last stanza:

Around four o'clock or so they begin to drift in.
The couple walking in the cemetery
where the stones flow from other centuries along the hills
notice how the silence gives way
to a few caws, and then more and more coast in
from somewhere, a steady, uneven stream
and a raucous chorus gathers in the trees.
The man sitting in the dentist chair
waiting for the dentist to appear, stares out the window
and sees the crows riding the air
descending onto the trees across the street,
a haunting sight he hadn't expected here.
And someone driving west through town is amazed
at the swirl of the flock across the winter sky,
hundreds, thousands, of black flecks across clouds
stirred with cold blazing light.
Wow, a natural wonder, he thinks,
the most beautiful thing he's ever seen in this city,
or maybe anywhere, and feels
it's a piece of luck to have crows in your city,
something to be grateful for,
to share the wintry earth with crows.

According to Norse mythology, the great god Odin had two ravens, named Thought and Memory. They flew around the world by day and at night came back to tell him what they saw. The god could not have picked better companions: little escapes the notice of the corvids. In Ireland, people still speak of having "raven's knowledge"—an oracular ability to see and know everything. In the classical period, members of the Roman College of Augurs were able to predict the future only because the events were foretold them by the voices of crows.

And perhaps this is the greatest blessing bestowed by the crows of Auburn: summoning both Thought and Memory, they may help us to remember to be grateful and inspire us to imagine a better future.

How *shall* we share the wintry earth with crows—and other birds?

At a time when we looked to the birds as teachers and seers, a recurrent tradition was that birds carry the souls of the dead to heaven. In *Don Quixote,* Miguel de Cervantes records a tradition that King Arthur took the form of a raven when he died, a bird-mediated miracle that would allow him one day to rule again. In China, it was widely believed that storks were the vehicle for the dead (interestingly, this tradition is reversed in the European tradition, showing storks bringing human babies into the world from the sky). Western seafaring nations held that at sea, gulls ferried the souls of sailors, and on land, ravens and crows conveyed all the others to paradise.

But birds were notably absent from that other land of the dead. In medieval times, biblical scholars imagined that Hell was a place with no birds. Today, at times, our kind seems determined to bring that Hell to earth. In their variety and sheer numbers, birds are among the most successful forms life has ever produced—yet today, one in eight bird species faces extinction because of human interference. BirdLife International, a global partnership of conservation organizations, reports that one quarter of all American bird species have declined since 1970. In the last four decades, the populations of twenty of our most beloved common birds have more than halved in number—including the northern bobwhite, who so sweetly but emphatically pronounces his own name, with a staggering population loss of 82 percent. In North and South America, twenty-eight species of hummingbird are threatened with extinction. In Europe, nearly half—45 percent—of common birds are declining. The familiar turtledove has lost 62 percent of its population in the last twenty-five years; the legendary nightingale is silently disappearing. Africa's magnificent raptors dwindle: in the past thirty years, eleven eagle species have declined by up to 98 percent in Mali, Niger, and Burkina Faso alone. Worldwide, 29 percent of all parrots are threatened with extinction. The list goes on and on.

At fault are the usual suspects: inhumane, industrial-scale factory farming and fishing usurp birds' habitat and pollute their food and water. Rapacious, wasteful logging destroys their forest homes. Humans bring foreign species to places they don't belong, displacing native species of plants and animals. Perhaps the biggest threat of all, according to the International Union for the Conservation of Nature, is global climate change, caused by our addiction to fossil fuels and to excess consumption of farmed meat. We claim to cherish birds for their songs, for their beauty, for their gift of flight. And yet, as we drive our cars and shop in our grocery stores and build our homes, we rob their food, usurp their nests, murder their young, and render their already arduous, miraculous migrations impossible.

Despite it all, a few bird species, like crows, manage to thrive. Yet as our own numbers swell—scientists note the recent chart of human population growth looks more like that of a disease bacterium than a vertebrate—we, in our hubris, regard these birds as pests. Once, the ancients recognized birds as our teachers, our guardians, our bridge to heaven. Today we need birds' blessings more than ever; more than ever we need to heed their message. Crows, in their great massing into urban roosts, remind us of something we must never forget. When Cornell's Kevin McGowan sees a roost of tens of thousands of crows, he thinks of this: "One hundred years ago, the passenger pigeon went extinct, and it was the most numerous bird on the planet. There were flocks of a billion birds. And when they flew by—it sounds like hyperbole, but it's not—a flock would darken the sky from horizon to horizon for hours." Once, their flocks stretched a mile wide and three hundred miles long, and the roar of their wings sounded like the winds of a hurricane. When they would alight, the branches of hardwood trees would crack beneath their weight. "Those were incredible congregations of birds," he says. "We shot them to death and we cut down the forest—and we drove the most abundant bird on earth to extinction.

"We'll never see flocks of passenger pigeons again," the ornithologist says. "We'll never see that incredible spectacle of wildlife. But we can still see large congregations of birds in some places. And to me, when I see a huge group of crows coming into town, making a lot of noise and flying in a big cloud, that's what I think of: I think, wow, I wonder what it would have been like to see passenger pigeons?"

By helping us recall those dazzling flocks, says McGowan, crows remind us of the ubiquity of birds' miracles. Birds are as ordinary as they are mysterious, as powerful as they are fragile, so like us and so beguilingly Other. Birds bring us the gifts of Thought and Memory, guided as they are both by intellect and instinct. These winged creatures, made of air, have outlived their kin, the dinosaurs. It is our duty and privilege to protect them.

Acknowledgments

Birds bring many blessings. Never before have I made so many wonderful new friends from researching and reporting a book.

Cassowaries brought the intrepid field biologist Dr. Andrew Mack into my life; African grey parrots introduced me to Dr. Irene Pepperberg and her tireless and knowledgeable lab manager, Arlene Levin; and they in turn introduced me to Snowball the dancing cockatoo and his delightful owner, Irena Schulz. To hummingbirds I owe a lasting friendship with bird rehabilitator and sculptor Brenda Sherburne and her family. Pigeons brought me to the Lesieur family. Hawks brought me to Nancy Cowan and her New Hampshire School of Falconry, where I still visit regularly. These are all people whose friendship I will cherish for the rest of my life.

I am also indebted to many old friends for their help with this book. Readers may have met some of the folks who appear in these pages before. From my memoir *The Good Good Pig*, readers will remember not only my hens, who begin this book, but also Selinda Chiquoine, our former tenant and my partner in my first foray into falconry; our neighbors Jarvis and Bobbie Coffin; and my friends Gretchen Vogel and Elizabeth Marshall Thomas. Our more recent tenant, Elizabeth Kenney, has a starring role in the chapter on chickens. I thank her husband, Nathan Townsend, too, for providing the wonderful picture of her surrounded by her Rangers.

For reading, commenting upon, and correcting errors in drafts of this manuscript, I thank Stephen Bodio (an author of exquisite books on both falcons and pigeons, who also introduced me to P.); Jarvis and Bobbie Coffin; Nancy Cowan; Joel Glick; Lita Judge; Elizabeth Kenney; Arlene Levin; my husband, writer and editor Howard Mansfield (who claims he doesn't even like birds all that much, but who has always cared tenderly for ours);

Howard Nelson (who also kindly gave permission to include three of his poems); Stacey O'Brien; Patty Park; Brenda Peterson; Eliza Schuster; Brenda Sherburn and her consort, Russ La Belle; Jody Simpson; Elizabeth Marshall Thomas; Gretchen Vogel; and scientists Richard Chipman of the USDA; Dr. Gary Galbreath of Northwestern University and the Field Museum in Chicago; Dr. Andrew Mack of the Carnegie Museum's Powdermill Nature Reserve; Dr. Aniruddh Patel of the Neurosciences Institute; Dr. Irene Pepperberg of Brandeis and Harvard; Dr. Irena Schulz of Bird Lovers Only; Dr. Charles Walcott of Cornell University; and Dr. David Westcott of the Australian Commonwealth Scientific and Industrial Research Organisation's Tropical Forest Research Centre. All errors that, despite their best efforts, may remain are, of course, mine.

In addition I thank Dianne Taylor-Snow, who helps me with every book; my friend and supporter Adele Carney; Sharon and Robin Wingrave, at whose Queensland home I stayed during the first days of searching for cassowaries, and whose medicine cabinet I emptied of Band-Aids; Frances Srulowitz, who introduced me to Brenda Sherburne; Kathy Schweitzer and her retinue of inspiring birds and beasts; Field Museum paleontologist Dr. Olivier Rieppel; Pennsylvania hummingbird rehabilitator Mary Birney and TriState Bird Rescue and Research Center; gracious Margaret Thorsborne; friend and neighbor, author and photographer Vicki Stiefel for the loan of her camera; Dave Judge for technological help; and budding "birdologist" Rev. Elaine Bomford.

I had the fabulous good luck to work on this with an editor who loves birds as much as I do: Leslie Meredith. (At our first meeting, she brought photos of her recent falconry lesson to show me over lunch.) Actually, our collaboration wasn't just good luck, but the result of the consistently excellent judgment of my literary agent, Sarah Jane Freymann. Thanks for making that perfect match.

Finally, I want to acknowledge three neighbors simply for making my life here in Hancock, New Hampshire, that much birdier: Don and Lillian Stokes, authors of the famous field guides, have graciously spent time with me and Howard, watching and narrating the adventures of loons and woodcocks and trying (valiantly, unsuccessfully) to educate us sufficiently about bluebird nest boxes to prevent our field from turning into a wren slum. Eric

Masterson shared with me the wonders of local owls and hawks. Once, he took me with him to witness the fall hawk migration over one of our mountains and said something I've always remembered: "The thing is," he told me as red-tails and broad-wings sailed high above our heads, "you're not just watching a bird. You're watching *life*."

Selected Bibliography

Chickens

Rossier, Jay. *Living with Chickens*. Guilford, Conn.: The Lyons Press, 2004.

Scientific articles:

Evans, C. S., and Linda Evans. "Chicken Food Calls Are Functionally Referential." *Animal Behavior* 58 (1999): 307–19.

Smith, C. L., and C. S. Evans. "Multi-Modal Signaling in Fowl, *Gallus gallus*." *Journal of Experimental Biology* 211 (2008): 2052–57.

Wilson, D., and C. S. Evans. "Mating Success Increases Alarm Calling in Male Fowl." *Animal Behavior* 76 (2008): 2029–35.

Website:

Cackle Hatchery Chicken catalog: www.cacklehatchery.com

Cassowary

Chatterjee, Sankar. *The Rise of Birds*. Baltimore: Johns Hopkins University Press, 1997.

Chiappe, Luis M. *Glorified Dinosaurs*. Sydney: University of New South Wales Press, 2007.

Fastovsky, David E., and David B. Weishampel. *The Evolution and Extinction of the Dinosaurs*, 2nd ed. New York: Cambridge University Press, 2005.

Fuller, Errol. *Extinct Birds*. Ithaca: Cornell University Press, 2001.

Manjep, Ian Saem, and Ralph Bulmer. *Birds of My Kalam Country*. Auckland: Auckland University Press, 1977.

Shipman, Pat. *Taking Wing: Archaeopteryx and the Evolution of Bird Flight*. New York: Simon & Schuster, 1998.

Scientific articles:

Cooper, Alan, Carles Lalueza-Fox, Simon Anderson, et al. "Complete Mitochondrial Genome Sequences of Two Extinct Moas Clarify Ratite Evolution." *Nature* 409 (2001): 704–7.

Crome, F. H. J., and L. S. Moore. "The Cassowary's Casque." *The Emu* 88 (1988): 123–24.

Harshman, J., E. L. Braun, M. J. Braun, et al. "Phylogenomic Evidence for Multiple Losses of Flight in Ratite Birds." *Proceedings of the National Academy of Sciences* 105 (2008): 13462–67.

Healey, Christopher J. "Pigs, Cassowaries and the Gift of the Flesh: A Symbolic Triad in Maring Cosmology." *Ethology* 24, no. 3 (July 1985): 153–65.

Kofron, Christopher P. "Attacks to Humans and Domestic Animals by the Southern Cassowary (*Casuarius casuarius johnsonii*) in Queensland, Australia. *Journal of Zoology* 249 (2000): 375–81.

Mack, Andrew L., and Gretchen Druliner. "A Non-Intrusive Method of Measuring Movements and Seed Dispersal in Cassowaries." *Journal of Field Ornithology* 74, no. 2 (2003): 193–96.

Mack, Andrew L., and Josh Jones. "Low-Frequency Vocalization by Cassowaries (*Casuarius ssp.*)." *The Auk* 120, no. 4 (2003): 1062–68.

Richardson, K. C. "The Bony Casque of the Southern Cassowary." *The Emu* 91 (1991): 56–58.

Slack, Kerryn E., Frederic Delsuc, Patricia Mclenachan, et al. "Resolving the Root of the Avian Mitogenomic Tree by Breaking Up Long Branches." *Molecular Phylogenetics and Evolution* 42 (2007): 1–13.

Westcott, D. A., and Devon L. Graham. "Patterns of Movement and Seed Dispersal of a Tropical Frugivore." *Oecologica* 122 (2000): 249–57.

Westcott, D. A., and K. E. Reid. "Use of Medetomidine for Capture and Restraint of Cassowaries (*Casuarius casuarius*)." *Australian Vet* 80, no. 3 (March 2002): 150–53.

Website:

Community for Coastal and Cassowary Conservation: www.cassowary.conservation .asn.au/

Hummingbirds

Newfield, Nancy L., and Barbara Nielsen. *Hummingbird Gardens.* Shelburne, Vt.: Chapters Publishing, 1996.

Tyrrell, Esther Quesada, and Robert A. Tyrrell. *Hummingbirds: Their Life and Behavior.* New York: Crown, 1985.

Williamson, Sheri L. *Hummingbirds of North America.* Boston: Houghton Mifflin, 2001.

Websites:

Brenda Sherburn's children's art cooperative for conservation: www.saveworlddraw .org

WildCare wild animal rehabilitation: www.wildcare.org

Hawks

Bodio, Stephen. *Eagle Dreams*. Guilford, Conn.: The Lyons Press, 2003.
————. *A Rage for Falcons*. Boulder, Colo.: Pruett Publishing Co., 1992.
Clark, William, and Brian K. Wheeler. *Hawks of North America*. Boston: Houghton Mifflin, 2001.
Ferguson-Lees, James, and David A. Christie. *Raptors of the World*. Boston: Houghton Mifflin, 2001.
Frederick II, Emperor of Hohenstaufen. *The Art of Falconry*. Translated by Casey A. Wood and F. Marjorie Fyfe. Stanford, Calif.: Stanford University Press, 1943.
Parry-Jones, Jemima. *Falconry*. Newton Abbot, Devon, England: David & Charles, 2003.

Website:

New Hampshire School of Falconry: www.nhschooloffalconry.com

Pigeons

Blechman, Andrew. *Pigeons: The Fascinating Saga of the World's Most Revered and Reviled Bird*. New York: Grove Press, 2006.
Bodio, Stephen. *Aloft: A Meditation on Pigeons and Pigeon-Flying*. New York: Lyons and Burford, 1990.
Gibbs, David, Eustace Barnes, and John Cox. *Pigeons and Doves*. New Haven: Yale University Press, 2001.
Humphries, Courtney. *Superdove: How the Pigeon Took Manhattan . . . and the World*. New York: HarperCollins, 2008.
Levi, Wendell M. *Encyclopedia of Pigeon Breeds*. Sumter, S.C.: Levi Publishing Co., 1996.

Scientific articles:

Heyers, D., Martina Manns, Harald Luksch, et al. "A Visual Pathway Links Brain Structures Active During Magnetic Compass Orientation in Migratory Birds." *PLos ONE* 2, no. 9 (2007): e937.
Jorge, Paulo E., Alice E. Marques, and John B. Phillips. "Activational Rather than Navigational Effects of Odors on Homing of Young Pigeons." *Current Biology* 19 (April 28, 2009): 650–54.
Thorup, Kasper, Isabelle-A. Bisson, Melissa S. Bowlin, et al. "Evidence for a Navigational Map Stretching Across the Continental U.S. in a Migratory Songbird." *PNAS* 104, no. 46 (November 2007): 18115–19.

Websites:

American Racing Pigeon Union: www.pigeon.org
Joe Black's Wings of Eternity White Dove Release: www.wingsofeternity.com
If you find a lost racing pigeon, e-mail this address: 911PigeonAlert-owner@yahoogroups.com

Parrots

Berger, Joanna. *The Parrot Who Owns Me*. New York: Random House, 2001.

Patel, Aniruddh D. *Music, Language, and the Brain*. New York: Oxford University Press, 2008.

Pepperberg, Irene M. *Alex & Me*. New York: HarperCollins, 2008.

———. *The Alex Studies: Cognitive and Communicative Abilities of Grey Parrots*. Cambridge: Harvard University Press, 1999.

Rothenberg, David. *Why Birds Sing*. New York: Basic Books, 2005.

Sacks, Oliver. *Musicophilia: Tales of Music and the Brain*. New York: Knopf, 2007.

Stap, Don. *Birdsong: A Natural History*. New York: Scribner, 2005.

Tweti, Mira. *Of Parrots and People*. New York: Viking, 2008.

Scientific articles:

Fehér, O., H. Wang, S. Saar, et al. "*De Novo* Establishment of Wild-type Song Culture in the Zebra Finch." *Nature* 459 (2009): 564–68.

Haesler, S., C. Rochefort, B. Georgi, et al. "Incomplete and Inaccurate Vocal Imitation after Knockdown of FoxP2 in Songbird Basal Ganglia Nucleus Area X." *PLoS Biol* 5, no. 12 (2007): e321.

Patel, Aniruddh D., John R. Iversen, Micah Bregman, and Irena Schulz. "Experimental Evidence for Synchronization to a Musical Beat in a Non-Human Animal." *Current Biology* 19 (May 26, 2009): 827–30.

Patel, Aniruddh D., John R. Iversen, Micah R. Bregman, et al. "Investigating the Human-Specificity of Synchronization to Music." *Proceedings of the Tenth International Conference of Music Perception and Cognition*, Sapporo, Japan, 2008, 100–103.

Schachner, Adena, Timothy F. Brady, Irene M. Pepperberg, and Marc D. Hauser. "Spontaneous Motor Entrainment to Music in Multiple Vocal Mimicking Species." *Current Biology* 19 (May 26, 2009): 831–36.

Websites:

The Alex Foundation: www.alexfoundation.org

Bird Lovers Only parrot rescue, with videos of Snowball dancing: www.birdloversonly.org

Crows

Heinrich, Bernd. *The Mind of the Raven*. New York: Harper Perennial, 2000.

Kilham, Lawrence. *The American Crow and Common Raven*. College Station: Texas A & M University Press, 1989.

Marzluff, John, and Tony Angell. *In the Company of Crows and Ravens*. New Haven: Yale University Press, 2005.

Savage, Candace. *Bird Brains*. San Francisco: Sierra Club Books, 1995.

Wilmore, Sylvia Bruce. *Crows, Jays, Ravens and Their Relatives*. Middlebury, Vt.: Paul Erikson, 1977.

Scientific articles:

Caffrey, Carolee, Shauna C. R. Smith, and Tiffany Weston. "West Nile Virus Devastates an American Crow Population." *The Condor* 107 (2005): 128–32.

Gorenzel, W. Paul, Terrell P. Salmon, Gary D. Simmons, et al. "Urban Crow Roosts—A Nationwide Phenomenon?" Wildlife Damage Management Conferences, Proceedings, University of Nebraska, 2000, 158–69.

Wier, Alex A. S., Jackie Chappell, and Alex Kacelnik. "Shaping of Hooks in New Caledonian Crows." *Science* 297, no. 5583 (August 9, 2002): 981.

Websites:

Trails.com: www.trails.com

Wild about Britain: www.wildaboutbritain.co.uk

To hear the author's radio report on the crows of Auburn: www.loe.org/shows/shows.htm?programID=06-P13-00003#feature7

General

Barber, Theodore Zenophon. *The Human Nature of Birds*. New York: St. Martin's Press, 1993.

Berger, Joanna. *Birds: A Visual Guide*. Buffalo, N.Y.: Firefly Books, 2006.

Dee, Tim. *A Year on the Wing*. New York: Free Press, 2009.

Gilbert, Suzie. *Flyaway: How a Wild Bird Rehabber Sought Adventure and Found Her Wings*. New York: HarperCollins, 2009.

Gill, Frank B. *Ornithology*, 2nd ed. New York: W. H. Freeman & Co., 1995.

Hill, Jen, ed. *An Exhilaration of Wings: The Literature of Birdwatching*. New York: Viking, 1999.

Kilham, Lawrence. *On Watching Birds*. Chelsea, Vt.: Chelsea Green, 1998.

Lorenz, Konrad. *King Solomon's Ring*. New York: Routledge, 2002.

———. *On Aggression*. San Diego: Harcourt Brace, 1996.

Skutch, Alexander E. *The Minds of Birds*. College Park: Texas A&M University Press, 1996.

Stokes, Donald, and Lillian Stokes. *Stokes Field Guide to the Birds: Eastern Region*. Boston: Little, Brown, 1996.

———. *Stokes Field Guide to the Birds: Western Region*. Boston: Little, Brown, 1996.

Tate, Peter. *Flights of Fancy: Birds in Myth, Legend, and Superstition*. New York: Delacorte Press, 2007.

Thomas, Elizabeth. *The Hidden Life of Deer*. New York: HarperCollins, 2009.

Tinbergen, Niko. *The Herring Gull's World*. New York: Basic Books, 1960.

Weidensaul, Scott. *The Ghost with Trembling Wings*. New York: North Point Press, 2002.

Williams, Terry Tempest. *Refuge: An Unnatural History of Family and Place*. New York: Vintage, 1992.

Index

Credits for the Photographs

About the Author

Sy Montgomery is a naturalist, author, documentary scriptwriter, and radio commentator who writes for children as well as adults. Among her award-winning books are *The Good Good Pig*, *Journey of the Pink Dolphins*, *Spell of the Tiger*, and *Search for the Golden Moon Bear*. She has made four trips to Peru and Brazil to study the pink dolphins of the Amazon. On other expeditions, she was chased by an angry silverback gorilla in Zaire; bitten by a vampire bat in Costa Rica; undressed by an orangutan in Borneo; and hunted by a tiger in India. She also worked in a pit crawling with eighteen thousand snakes in Manitoba; handled a wild tarantula in French Guiana; and swam with piranhas, electric eels, and dolphins in the Amazon. She lives in New Hampshire with her husband, border collie, and chickens.